Powered by Enhanced WebA

Workbook for the Accuplacer® and Compass® Mathematics Exam

Edward L. Green

John Jay College of Criminal Justice

Borough of Manhattan Community College

Pace University

BROOKS/COLE
CENGAGE Learning·

Australia · Brazil · Japan · Korea · Mexico · Singapore · Spain · United Kingdom · United States

BROOKS/COLE
CENGAGE Learning·

Workbook for the Accuplacer® and Compass® Mathematics Exam: Powered by Enhanced WebAssign
Edward L. Green

Publisher: Charlie Van Wagner

Acquisitions Editor: Marc Bove

Developmental Editors: Stefanie Beeck, Donald Gecewicz

Assistant Editor: Shaun Williams

Editorial Assistant: Carrie Jones

Media Editor: Bryon P. Spencer

Marketing Manager: Gordon Lee

Marketing Assistant: Shannon Maier

Marketing Communications Manager: Darlene Macanan

Content Project Manager: Carol Samet

Creative Director: Rob Hugel

Art Director: Vernon Boes

Print Buyer: Becky Cross

Production Service: Scratchgravel Publishing Services

Text Designer: Vernon Boes

Copy Editor: Carol Lombardi

Cover Designer: Irene Morris

Cover Image: PhotoAlto/Alix Minde/Getty Images

Compositor: Scratchgravel Publishing Services

For product information and technology assistance, contact us at
Cengage Learning Customer & Sales Support, 1-800-354-9706.

For permission to use material from this text or product, submit all requests online at **www.cengage.com/permissions**.
Further permissions questions can be emailed to **permissionrequest@cengage.com**.

Library of Congress Catalog Number: 2011939518

ISBN-13: 978-1-133-11361-4
ISBN-10: 1-133-11361-3

Brooks/Cole
20 Davis Drive
Belmont, CA 94002-3098
USA

Cengage Learning is a leading provider of customized learning solutions with office locations around the globe, including Singapore, the United Kingdom, Australia, Mexico, Brazil, and Japan. Locate your local office at: **www.cengage.com/global**.

Cengage Learning products are represented in Canada by Nelson Education, Ltd.

To learn more about Brooks/Cole, visit
www.cengage.com/brookscole.

Purchase any of our products at your local college store or at our preferred online store **www.CengageBrain.com**.

Printed in the United States of America
1 2 3 4 5 6 7 15 14 13 12 11

This book is dedicated to my fantastic and wonderful grandchildren,
Isabella Morgan, Sivon Michelle, Emma Reese,
Aaron Jack, and Dylan Chase

Contents

Preface

I have long been aware of a serious gap in the mathematics textbooks used by high school and college classes: Students arrive at their first college mathematics class proficient in basic arithmetic, algebra, and geometry skills, but are unable to pass the Accuplacer® or Compass® exams.

I wrote this book to fill that gap.

The *Workbook for Accuplacer® and Compass® Mathematics Exam* focuses on step-by-step processes and critical thinking needed to solve complex problems requiring multiple steps and formulas:

- Most chapters begin with a list of Key Terms, so fundamental concepts are easy to find and refer to.

- Students are guided through the process with clear, concise instructions. Examples are given to ensure quick recall and recognition.

- Exercises thoroughly test the students' grasp of the concepts presented and instill confidence as each exercise set is completed.

- The chapters progress in logical order from basic to complex mathematical concepts.

- Practice Exams at the end of the text provide useful review and practice for the real thing.

Equally important is the one-page "You *Can* Learn Mathematics!" presented on page xi. Many students struggle in college mathematics classes because of previous difficulty, anxiety, or lack of academic support. This brief list of strategies serves as both guideline and wake-up call to help students achieve success—not only in mathematics, but also in other endeavors.

The *Workbook for Accuplacer® and Compass® Mathematics Exam* is intended to help students review concepts they previously learned before their instructors administer the placement exams. It can be used as review for students to practice the basics on their own as they progress through their Prealgebra and College Algebra courses. If they have access to Enhanced WebAssign, they can refer to the workbook and practice the basics as they navigate through their courses. In addition, instructors can use this workbook to provide students with more practice problems in any of their algebra courses, to help them relearn the skills they might have forgotten.

To the Student

Stop the negatives

"I cannot learn mathematics." . . . "I always fail my mathematics tests."

Be positive

"I can learn mathematics." . . . "I will pass my mathematics tests." . . . "I will be successful."

It is not all your fault. Many high schools do not teach college-level mathematics. Thus, you have not learned the necessary skills needed to take college credit mathematics classes. This stops at the college level.

The Compass® and Accuplacer® Examinations are multiple-choice, untimed exams taken on a computer. The mathematics component is an adoptive test: This means that the next question will depend upon your success on the previous question. No two students will take exactly the same exam. You see one question at a time and cannot see the next question until the previous question is answered.

Traditional texts might enable you to become proficient in basic skills in arithmetic, algebra, and geometry. Yet you might not be able to pass the Compass® and Accuplacer® Exams. There is a gap in your learning. You have not developed the critical thinking skills needed to solve problems that require a multitude of steps and formulas.

This *Workbook for Accuplacer® and Compass® Mathematics Exam* fills this gap by giving you exercises similar to those found on the Compass® and Accuplacer® Exams. After mastering the basic skills in arithmetic, algebra, and geometry, you can then focus on the Enhanced WebAssign homework.

If an obstacle occurs and you cannot solve a problem, do not become negative. There is help. Ask your professor or go to the math resource center at your college. Seek tutoring immediately. You can be a success in mathematics!

Acknowledgments

This book is dedicated to the encouragement and support my family has given me through the years. During my childhood, my mother, Fae Sarah Wigder Green, encouraged me to pursue my education while my father, Jack Green, worked long hours to support his family. My brother, Dr. William Green, always helped me to excel in school, especially in mathematics and science. As a Diplomate in Rehabilitation Medicine, he is the recipient of numerous medical awards for outstanding work. Furthermore, my aunts, Rose Saldinger, Molly Wigder, and Ida Wigder, assisted me to achieve my goals. My brother-in-law, José Chavin Pressner, is a poet and a professor of language. He gave me technical assistance.

I am very fortunate in that my beautiful wife, Zelma, always encouraged me to pursue my career. While I was working late at night, she graduated from Brooklyn College with her Master's Degree in Spanish Literature while taking excellent care of our three wonderful sons Elliot, Philip Steven, and Seth Andrew. Now, she is an Adjunct Assistant Professor at Pace University in the Department of Modern Languages and Cultures.

My three sons married intelligent women who encourage their husbands to pursue their dreams. My daughters-in-law Holly, Edit, and Randi are like daughters to me. I am a lucky grandfather with five grandchildren: Isabella Morgan, Sivon Michelle, Emma Reese, Aaron Jack, and Dylan Chase. I always encourage them to be the best they can.

I would like to thank the following people at Cengage Learning, who encouraged and supported me as I completed this project: Charlie Van Wagner, Marc Bove, Stefanie Beeck, Donald Gecewicz, and the rest of the publishing team.

You *Can* Learn Mathematics!

The right attitude is necessary to be successful in mathematics:

1. Begin with an open mind.
2. Overcome low self-esteem.
3. Overcome a weak mathematics background.
4. Do not make excuses.
5. Take a positive approach.
6. Make an exceptional effort from the very first day.

Use your class time effectively:

1. Learn as much as you can during class.
2. Attend all classes and be punctual.
3. Organize your notebook.
4. Take complete class notes, and if you miss some material, ask a fellow student.
5. Ask questions about material you do not understand. No question is stupid. You are paying for this course.
6. Get help outside the class. Use your school's Mathematics Resource Center.

Get the maximum benefit out of studying:

1. Study your class notes. Review the sample problems in the textbook before starting the Enhanced WebAssign homework assignments.
2. Do your homework assignments before the next class.
3. Constantly review previously learned concepts.
4. Work with other students. You can all benefit from the interaction.
5. Prepare questions to ask during the next class.

Overcome math anxiety:

1. Have a positive attitude toward learning mathematics. This attitude will build self-confidence and reduce math anxiety. Forget past negatives.
2. Mathematics is a cumulative subject. Seek immediate tutoring if you do not understand the topics being taught.
3. Constant studying, including developing study groups and visiting the math tutoring center, will reduce math anxiety and build self-esteem. Make flash cards of formulas and vocabulary.
4. Do all homework assignments on time, attend all classes, be punctual, and take complete notes.
5. Read your math text, study the sample problems, review your notes, and practice the Enhanced WebAssign homework. Write down the problems you find difficult, and seek immediate tutorial help. Do not let the difficulties accumulate.
6. Do not be embarrassed to ask questions. No question is stupid. Asking questions will help you overcome math anxiety by increasing your conceptual knowledge and reducing your frustration.

Prepare thoroughly for Compass® and Accuplacer® Exams:

1. Understand the causes of math anxiety.

2. Avoid math anxiety with preparation.

3. Study all the mathematics concepts before using the *Compass®-Accuplacer® Study Guide* or the Enhanced WebAssign homework.

4. Do not rush. It is an untimed exam. Before placing the correct answer in the portal, do the problem again. You cannot go back and change your previous answer.

5. Prepare materials needed for the test at least one day before.

6. Arrive early to the test site.

7. Use those test strategies that your professor emphasized. For example, eliminate the answers you know are wrong.

8. Read the problem slowly and carefully. List the information given and what you have to find. Outline a step-by-step approach to solve the problem.

9. Before doing the Enhanced WebAssign homework, read over the material, study the concepts, and then do the sample problems.

PART I

ARITHMETIC

Chapter 1

Numerals

Key Terms

Integer An integer is a whole number—not a fraction—that may be positive or negative.

The following table illustrates how the positions of integers and zeros indicate value.

1,000,000s	100,000s	10,000s	1000s	100s	10s	1s
Millions	Hundred thousands	Ten thousands	Thousands	Hundreds	Tens	Ones

Exercises

Write the correct answers on the answer lines.

1. Fifty-four thousand three hundred is written: _____

   a. 54,030

   b. 540,300

   c. 54,300

   d. 54,000,300

   e. 54,003

2. Three million twelve is written: _____

   a. 30,000,012

   b. 3,000,120

   c. 30,012

   d. 300,000,012

   e. 3,000,012

3. Ten million one hundred sixty-three is written: _____

    a. 10,000,163

    b. 1,000,163

    c. 100,163

    d. 100,000,163

    e. 100,000,063

4. Write in words: 6,400,005 _____

_____

5. Write in words: 325,109 _____

_____

6. Write in words: 6,000,195,071 _____

_____

7. Which one of the following is the largest? _____

    a. 3,400,325

    b. 3,400,025

    c. 3,400,425

    d. 3,400,427

    e. 3,400,421

8. Which one of the following is the smallest? _____

    a. 3,600,025

    b. 3,610,025

    c. 3,600,029

    d. 3,610,029

    e. 3,600,028

9. Which one of the following is the largest? _____

    a. 435,001

    b. 395,001

    c. 436,001

    d. 436,007

    e. 436,107

10. Which one of the following is the smallest? _____

    a. 395,001

    b. 328,302

    c. 428,009

    d. 329,401

    e. 328,102

11. Write as a numeral: nine billion twenty thousand three. _____

12. Write as a numeral: three million one hundred six. _____

Chapter 2

Adding and Subtracting Units

The following information will help you answer the questions in this section:

1 hour	=	60 minutes
1 minute	=	60 seconds
1 pound	=	16 ounces
1 yard	=	3 feet
1 foot	=	12 inches

Exercises

Write the correct answers in the spaces provided.

1. A movie starts at 7:45 p.m. and ends at 9:05 p.m. How long is the movie?

   _____

2. Ryoko and her friend Tommy saw two movies. The first movie was 1 hour and 35 minutes, and the second movie was 1 hour and 49 minutes. How long were both movies?

   _____

3. A lumberjack cuts three pieces of wood: One is 3 feet long, one is 7 inches long, and one is 2 feet 9 inches long. How much wood did he cut?

   _____

4. Tamara cuts three pieces of wood from a 9-yard board. The first piece was 2 feet 7 inches, the second piece was 1 foot 9 inches, and the third piece was 7 inches. How much wood was left over?

   _____

5. A flight left Kennedy Airport at 6:42 p.m. and arrived in Miami at 10:19 p.m. How long did the flight last?

   _____

6. Danya made 10 dresses. Each dress requires 1 yard and 2 feet of material. How much material is left over from a 22-yard-long roll?

   _____

7. One pound of ground sirloin beef costs $4.80. What is the cost of 1 pound 3 ounces?

   _____

8. Empire Turkey costs 89 cents a pound. What is the cost of a 14-pound 7-ounce turkey?

   _____

9. Two pounds of pastrami cost $18.00. What is the cost of 3 ounces?

   _____

10. Math 103 class begins at 3:35 p.m. and ends at 5:05 p.m. How long is the class?

   _____

11. While crating apples for supermarket distribution, an inspector discards spoiled apples. On Monday, the inspector found that 6% of the 66,000 pounds of apples were spoiled. How many pounds of apples were not spoiled?

_____

12. What is the maximum number of pieces of cloth $7\frac{1}{2}$ feet long that can be cut from a 70-yard roll of material?

_____

13. A garment worker has a 175-foot roll of cloth. She cuts 9 pieces $3\frac{1}{2}$ feet long, 6 pieces $5\frac{1}{4}$ feet long, and 7 pieces 3 yards long. How much cloth is left over?

_____

14. Rufus is putting up shelves in his new apartment. He is cutting $4\frac{1}{4}$ -foot pieces from a board 20 feet long. If the saw removes $\frac{1}{4}$ foot with every cut, how much of the board is left over after he makes the maximum number of cuts?

_____

   a. 3 feet

   b. 2 feet

   c. 1 foot

   d. 4 feet

   e. $\frac{1}{2}$ foot

15. A photographer wants to enlarge the length and width of a picture by $\frac{2}{5}$. If the original picture is 20 inches by 10 inches, what will be the dimensions, in inches, of the enlarged picture?

_____

   a. 28 inches by 14 inches

   b. 24 inches by 12 inches

   c. 12 inches by 9 inches

   d. 14 inches by 7 inches

   e. 22 inches by 11 inches

16. To decorate baskets for a wedding, Tamara needs the following amounts of ribbon for each basket:

Number of Ribbons	Length in Inches
7	6
5	14
4	14

If the ribbon costs $0.35 per foot, which of the following is the approximate cost of ribbon for all 11 baskets? _____

a. $4.90

b. $51.90

c. $54.00

d. $49.90

e. $53.90

17. The cost of electricity in Newark is $0.09 for each of the first 600 kilowatt-hours and $0.12 for each kilowatt-hour over 600. Find the number of kilowatt-hours used by Brigida if she receives a bill for $78.00. _____

a. 800

b. 700

c. 672

d. 900

e. 784

18. John cuts six pieces of wood, each $1\frac{1}{6}$ feet long, and four pieces of wood, each $2\frac{1}{4}$ feet long. If these pieces are cut from a 9-yard length of wood, how much wood remains? _____

Chapter 3

Fractions

Key Terms

Denominator   The bottom number of a fraction, which tells the number of equal parts in a fraction. It cannot be zero.

Factors   Two or more numbers that, when multiplied, give you a product.

Fraction   It is part of a whole. The denominator cannot equal zero.

Improper Fraction   A fraction in which the numerator is equal to or greater than the denominator.

Least Common Multiple (LCM)   The least multiple (excluding zero) of two or more numbers. For example, the LCM of 6 and 10 is 30.

Numerator   The top number in a fraction, which tells how many equal parts of the whole are being considered.

Proper Fraction   A fraction in which the numerator is less than the denominator.

Reducing Fractions

To reduce a proper fraction to lowest terms, find the largest number that divides evenly into the numerator and into the denominator.

To reduce an improper fraction (in which the numerator is equal to or larger than the denominator) to lowest terms, divide the denominator into the numerator and then see if the fraction part of your answer can be further reduced.

Addition of Fractions

1. The fractions being added must have a common denominator.
2. If the denominators are different, find the lowest common denominator and re-evaluate the numerators to find equivalent fractions.

3. Add numerators and rewrite the common denominator. Also add the integral parts of the fractions if necessary.

4. Reduce answer to lowest terms.

Subtraction of Fractions

1. Fractions must have a common denominator.

2. If denominators are different, find the lowest common denominator and re-evaluate the numerators to find equivalent fractions.

3. If the first fraction is smaller than the second fraction, borrow 1 from the integral part in the form of a fraction.

4. Subtract numerators and rewrite the common denominator. Also subtract the integral parts of the fractions if necessary.

5. Reduce answer to lowest terms.

Multiplication of Fractions

1. Change mixed numbers to improper fractions.

2. Cancel by reducing the numerators and denominators by the same factor.

3. Multiply the numerators and then multiply the denominators.

4. Reduce answer to lowest terms.

Division of Fractions

1. Change mixed numbers to improper fractions.

2. Invert (turn upside down) the fraction after the division sign.

3. Change the division sign to a multiplication sign.

4. Follow the rules for multiplication. Cancel and then multiply numerators and denominators.

5. Reduce answer to lowest terms.

Comparing Sizes of Fractions

1. Find the equivalent fraction if denominators are different.

2. Compare the numerators. The fraction with the largest numerator is the largest. The fraction with the smallest numerator is the smallest.

3. Sometimes it is easier to cross-multiply and to keep the largest or smallest fractions. Continue this process until the *largest* or *smallest* fraction is found.

Write the correct answers in the spaces provided.

1. If $\left(\dfrac{5}{6} - \dfrac{1}{4}\right) + \left(\dfrac{2}{5} + \dfrac{1}{20}\right)$ is calculated and the answer is reduced to lowest

   terms, what is the denominator of the resulting fraction? _____

   a. 30

   b. 10

   c. 15

   d. 20

   e. 25

2. $\left(\dfrac{3}{5} \div \dfrac{5}{3}\right) + \dfrac{4}{5} - \left(\dfrac{5}{6} \cdot \dfrac{6}{5}\right) = \,?$ _____

   a. $\dfrac{4}{5}$

   b. $\dfrac{4}{25}$

   c. $\dfrac{5}{4}$

   d. $\dfrac{25}{4}$

   e. $\dfrac{5}{25}$

3. Simplify: $\dfrac{4}{7} - \dfrac{3}{14} + 9$ _____

4. Simplify: $\dfrac{\dfrac{4}{5} - \dfrac{1}{2}}{\dfrac{3}{4}}$ _____

5. Simplify: $\dfrac{1}{4} + \left(\dfrac{3}{7} \cdot \dfrac{21}{6}\right) + \left(\dfrac{7}{9} \div \dfrac{1}{3}\right)$ _____

6. Simplify: $\dfrac{1}{2} + \left(\dfrac{3}{5} \cdot \dfrac{5}{9} \right) - \left(\dfrac{7}{12} \div \dfrac{1}{2} \right)$ _____

7. Simplify: $\left(\dfrac{3}{9} - 2 \right) \div \left(\dfrac{3}{8} \div \dfrac{1}{2} \right)$ _____

8. Simplify: $6 \div \dfrac{4}{9} - \dfrac{1}{18}$ _____

9. Simplify: $\dfrac{-\dfrac{2}{3}}{\dfrac{1}{2} - \dfrac{5}{6}}$ _____

10. Which of the following is smaller than $\dfrac{2}{7}$? _____

    a. $\dfrac{4}{9}$

    b. $\dfrac{1}{4}$

    c. $\dfrac{3}{14}$

    d. $\dfrac{5}{9}$

11. Which of the following is larger than $\dfrac{5}{9}$? _____

    a. $\dfrac{5}{8}$

    b. $\dfrac{3}{7}$

    c. $\dfrac{2}{9}$

    d. $\dfrac{11}{18}$

12. Which of the following fractions is the smallest? _____

    a. $\dfrac{2}{7}$

    b. $\dfrac{1}{9}$

    c. $\dfrac{3}{4}$

    d. $\dfrac{11}{15}$

    e. $\dfrac{4}{11}$

13. Which of the following fractions is the largest? _____

    a. $\dfrac{3}{7}$

    b. $\dfrac{2}{9}$

    c. $\dfrac{2}{3}$

    d. $\dfrac{3}{5}$

    e. $\dfrac{11}{15}$

14. Which of the following fractions is the largest? $\dfrac{2}{5}, \dfrac{3}{7}, \dfrac{4}{9}, \dfrac{3}{8}$ _____

15. Which of the following fractions is the smallest? $\dfrac{3}{7}, \dfrac{2}{9}, \dfrac{1}{5}, \dfrac{5}{8}$ _____

16. Nancy is buying candy for her daughter's birthday party. She buys 4 pounds of candy bars, 5 pounds of chocolate candy, and 3 pounds of jelly candy fish. If each child is given $\dfrac{1}{4}$ pound of candy, how many children can she feed at the birthday party?

_____

17. A recipe requires $\frac{3}{4}$ cup of tomato pulp. Which of the following calculations gives the number of cups of tomato pulp that should be used to make $\frac{1}{3}$ of the recipe?

   a. $\frac{3}{4} - \frac{1}{3}$

   b. $\frac{3}{4} + \frac{1}{3}$

   c. $\frac{3}{4} \cdot \frac{1}{3}$

   d. $\frac{3}{4} \div \frac{1}{3}$

   e. $\frac{1}{3} \div \frac{3}{4}$

18. $\frac{5}{6} + \left(\frac{4}{5} \div \frac{3}{4}\right) - \left(\frac{2}{3} \cdot \frac{1}{2}\right) = ?$

   a. $\frac{1}{17}$

   b. $\frac{19}{29}$

   c. $\frac{18}{13}$

   d. $\frac{41}{18}$

   e. $1\frac{17}{30}$

19. Rodriguez, Jones, and Schwartz decided to divide their winnings from a lottery ticket. Rodriguez receives $\frac{1}{6}$ of the winnings, Jones receives $\frac{5}{12}$ of the winnings, and the rest is received by Schwartz. What fraction of the winnings does Schwartz receive?

20. In a three-way contest for state senate, Candidate A received $\frac{1}{4}$ of the votes and Candidate B received $\frac{2}{3}$ of the votes. If 60,000 votes were counted, how many votes did Candidate C receive?

21. On a 40-question mathematics exam, Chitra answered 38 questions correctly. What fraction of the questions did she get wrong?   _____

22. At a town hall meeting, 14 men and 18 women attended. What fraction of the people who attended were women?   _____

Chapter 4

Decimals

Key Terms

Decimal A numeral that uses place value digits and a decimal point to write numbers that show tenths, hundredths, thousandths, etc.

Tenth One of ten equal parts (one tenth $= \dfrac{1}{10}$ or 0.1)

Hundredth One of a hundred equal parts (one hundredth $= \dfrac{1}{100} = 0.01$)

Thousandth One of a thousand equal parts (one thousandth $= \dfrac{1}{1000} = 0.001$)

Place Values

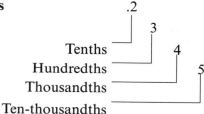

Comparing the Sizes of Decimals

1. First compare the integral parts.

2. If the integral parts are the same, compare the first numbers after the decimal point (tenths digits).

3. If the tenths digits are the same, then compare the second numbers after the decimal point (the hundredths digits).

4. Continue this process until the correct answer is found.

Addition and Subtraction of Decimals

1. Line up decimal points.

2. Add or subtract in appropriate columns.

3. Carry down the decimal point.

Multiplication of Decimals

1. Multiply as an ordinary problem and obtain the product.

2. Determine where the decimal point belongs in the product by counting the total number of decimal digits after the decimal point in both numbers being multiplied.

3. Place the decimal point in the product by counting the total number of places from Step 2 from right to left. If there are not enough digits, then add the appropriate number of zeros.

Division of Decimals

1. The number after the division sign is called the divisor. The number before the division sign is called the dividend. The answer to a division problem is called the quotient.

2. Move the decimal point as many places as necessary to make the divisor an integer.

3. Move the decimal point in the dividend the same number of places. If there are not enough places, the appropriate number of zeros must be added.

4. Proceed to divide. Write the decimal point in the correct place in the quotient. It should be in the same place as in the dividend.

Exercises

Write the correct answers in the spaces provided.

1. Name the place value of the digit 7 in this number: 68.097815 _____

   a. ten-thousandths

   b. hundred-thousandths

   c. millionths

   d. thousandths

   e. hundredths

2. Name the place value of the digit 9 in this number: 68.097815 _____

3. Name the place value of the digit 8 in this number: 62.097815 _____

4. Add: 53.838 + 2.87 + 2.763 _____

   a. 59.471

   b. 59.369

   c. 59.358

   d. 69.572

   e. 64.582

5. Add: 87.1 + 4.008 + 977.22 _____

   a. 1068.328

   b. 1058.228

   c. 1067.218

   d. 1068.118

   e. 1068.348

6. Add: 10.757 + 6.21 + 71.412 _____

   a. 88.378

   b. 88.379

   c. 760

   d. 76.000

   e. 88.48

7. Multiply: $7.6 \times 10,000$ _____

   a. 76

   b. 7600

   c. 760

   d. 76,000

   e. 760,000

8. Divide: $64,029 \div 1,000$ _____

   a. 64,029

   b. 6.4029

   c. 640.29

   d. 6402.9

   e. 64.029

9. Multiply: $6.66 \times 10,000$ _____

    a. 66,600

    b. 6660

    c. 666

    d. 66.6

    e. 666,000

10. What is the word name for 0.055? _____

    a. fifty-five hundredths

    b. fifty-five hundreds

    c. fifty-five thousands

    d. fifty-five thousandths

    e. None of the above

11. Subtract: $6.2 - 3.879$ _____

12. Subtract: $6 - 0.975$ _____

13. Divide: $0.02 \div 2$ _____

14. Divide: $2 \div 0.02$ _____

15. Multiply: 0.01×0.01 _____

16. Multiply: 0.2×0.019 _____

17. Mrs. Rodriguez went to the supermarket to buy supplies for the month. She bought 3.45 pounds of sugar, $2\frac{1}{3}$ pounds of carrots, and $5\frac{2}{3}$ pounds of corn. How many pounds do these items weigh? _____

    a. 9.45

    b. 10.61

    c. 10.45

    d. 11.45

    e. 11.35

18. Harris Rental Car charges $15 a day and $0.20 per mile for renting a car. You rented a car for 4 days and drove 256 miles. Find the total cost of renting the car. _____

    a. $66.12

    b. $66.20

    c. $65.12

    d. $51.20

    e. $111.20

19. Tamika is making toys to sell. Each toy costs her $1.70 to make. If she sells the toys for $2.90 each, how many will she have to sell to make a profit of exactly $36.00? _____

20. Maria bought meat for the Labor Day barbecue. She purchased 24.2 pounds of ground shoulder steak and 17.6 pounds of chicken. If 75 people ate all the food, how much on average did each person eat? _____

21. Jamal buys 1400 CDs for his music store. He pays $5.95 for each CD. He sells the first 900 for $13.50 each, and the rest for $9.25 each. What was his profit on the shipment?

_____

22. John buys an electronic toy for $16.80. He sells it on the Internet for $23.90. How many must he sell to make a profit of at least $320?

_____

23. An electric motor costing $445.21 has an operating cost of $0.023 for 1 hour of operation. Find the cost to buy and run the motor for 60 hours.

_____

   a. $0.14

   b. $13.80

   c. $1.79

   d. $446.59

   e. $444.91

24. Quan bought a car for $3500 down and made payments of $395.77 for 48 months. Find (1) the amount of the payments over the 48 months and (2) the total cost of the car.

_____

   a. (1) $18,996.96; (2) $22,871.60

   b. (1) $17,954.51; (2) $22,871.60

   c. (1) $19,735.31; (2) $22,979.08

   d. (1) $18,996.96; (2) $22,496.96

   e. (1) $17,954.51; (2) $22,496.96

25. Naomi Thompson owned 399.739 shares of a mutual fund on January 1. On December 31 of the same year, she had 487.661 shares. What was the increase in the number of shares Naomi owned during the year?

_____

   a. 66.922

   b. 197.922

   c. 76.922

   d. 87.922

   e. None of the above

26. Write these quantities in order of decreasing size. 3.333, 3.099, 3.1, 4.01 _____

27. Write these quantities in order of increasing size. 3.333, 3.099, 3.1, 4.01 _____

28. Which of the following is the smallest? _____

    a. 5.09

    b. 0.059

    c. 5.9

    d. 5.009

    e. 5.097

29. Which of the following is the largest? _____

    a. 0.88

    b. 0.0881

    c. 0.0808

    d. 0.8088

    e. 0.0880

30. Which number is the smallest? _____

    a. 0.62

    b. 0.6

    c. 0.65

    d. 0.69

    e. 0.605

31. Which number is the largest? _____

    a. 0.4562

    b. 0.451

    c. 0.452

    d. 0.457

    e. 0.458

32. Dinner for two cost $39.47. Mai left a $5.50 tip, and the tax was $3.79. She gave the waitress a $100 bill. How much change did she receive?

_____

33. Gas cost $3.80 per gallon. Rajiv drove 140 miles to his girlfriend's house at a rate of 20 miles per gallon. How much money did the trip cost?

_____

34. John buys 70 feet of ribbon for a party. He cuts 6 ribbons, each 3.75 feet long, and 7 ribbons, each 4.25 feet long. How many feet of ribbon does he have left over?

_____

35. From a 60-foot length of cable, Ichiro cuts 6 pieces: 1.2 feet, 0.6 foot, 3.2 feet, 7 feet, 2.85 feet, and 9.75 feet. How much cable is left over?

_____

36. A truck can hold 0.75 tons of bricks. In order to deliver 8.75 tons of bricks to the construction site, how many truckloads are needed?

_____

37. In Chicago, a teacher's salary is $44,400 per year. The school year is 200 days. If the school year were extended to 220 days, what annual salary should the teacher be paid if the salary is based on the same daily rate?

_____

  a. $47,840

  b. $39,960

  c. $48,840

  d. $48,900

  e. $38,960

38. You are coating your rectangular driveway with blacktop sealer. The driveway is 100.1 feet long and 40.1 feet wide. One can of sealer costs $50 and covers 250 square feet. How much will it cost to coat your driveway? _____

    a. $600

    b. $1200

    c. $1600

    d. $800

    e. $400

39. Mr. Johnson leaves his two children $5000. The oldest receives 0.50 more than the youngest. How much does the oldest receive? _____

    a. $3500

    b. $2500

    c. $1500

    d. $3000

    e. $2000

Chapter 5

Conversions: Decimals, Fractions, and Percents

Convert a Decimal to a Fraction

Rewrite the decimal as a fraction and reduce the answer to lowest terms.

The amount of numbers after the decimal point determines the number of zeros in the denominator, as follows:

- If there is only one number after the decimal point, the denominator in the fraction is 10.

- If there are two numbers after the decimal point, the denominator is 100.

- If there are three numbers after the decimal point, the denominator is 1000.

Convert a Percent to a Fraction

1. Cross out the percent symbol.

2. Create a fraction by using 100 as the denominator.

3. Reduce answer to lowest terms.

Convert a Percent to a Decimal

1. Place decimal point to the left of the percent symbol.

2. Cross out the percent symbol.

3. Move the decimal point two digits to the left.

Convert a Fraction to a Decimal

To change a fraction to a decimal, divide the denominator into the numerator. To do this, add a decimal point after the denominator and, after the decimal point, enough zeros (at least two) to complete the division.

Convert a Decimal to a Percent

Multiply by 100 (that is, move decimal point two places to the right).

Convert a Fraction to a Percent

Change the fraction to a decimal as described above. Multiply decimal by 100 (move decimal point two places to the right).

Write the correct answers in the spaces provided.

1. Express $\dfrac{9}{25}$ as a percent (write remainder in fractional form). _____

2. Express $\dfrac{7}{20}$ as a percent. _____

3. Express $5\dfrac{1}{2}\%$ as a fraction. _____

4. Express 2.125 as a percent. _____

5. Express 3.7% as a decimal. _____

6. Express 0.7 as a percent. _____

7. Write 0.008 as a fraction in lowest terms.   _____

8. Write 135% as a fraction reduced to lowest terms.   _____

9. Write this decimal as a fraction: 0.419   _____

    a. $\dfrac{419}{10}$

    b. $\dfrac{419}{1000}$

    c. $\dfrac{419}{100}$

    d. $4\dfrac{19}{100}$

    e. $4\dfrac{19}{1000}$

10. Write this fraction as a decimal: $\dfrac{25}{10,000}$ _____

    a. 25

    b. 0.0025

    c. 2.5

    d. 0.025

    e. None of the above

11. Express 7.85 as a fraction.   _____

12. Express 0.0345 as a percent.   _____

13. Write 0.0364 as a percent.   _____

14. Express $3\frac{7}{9}$ as a percent (remainder to be written in fractional form). _____

15. Write $\frac{4}{25}$ in decimal notation. _____

16. Express $\frac{6}{30}$ as a percent. _____

17. Express $\frac{7}{20}$ as a decimal. _____

18. Express 0.35 as a fraction. _____

19. Express 0.02 as a fraction. _____

20. Express 37% as a fraction. _____

21. Express 145% as a fraction. _____

22. Write 39% in decimal notation. _____

23. Write 145% in decimal notation. _____

24. If the fractions $\frac{3}{4}$, $\frac{2}{5}$, $\frac{8}{2}$, and $\frac{3}{12}$ are changed to decimals, and the sum of the first two numbers is subtracted from the sum of the last two numbers, what is the result? _____

Chapter 6

Rounding Off

Round to Tenths

To round to the nearest *tenth*, the answer must be expressed to only *one* decimal digit (one number to the right of the decimal point). Look at the second decimal digit (second number to the right of the decimal point). If the second digit is 5 or more, the first decimal digit must be made one digit larger (examples: 0.69 = 0.7 or 0.61 = 0.6).

Round to Hundredths

To round to the nearest *hundredth*, the answer must be expressed to only *two* decimal digits (two numbers to the right of the decimal point). Look at the third decimal digit (third number to the right of the decimal point). If the third digit is 5 or more, the second decimal digit must be made one digit larger (examples: 0.576 = 0.58 or 0.574 = 0.57).

Round to Thousandths

To round to the nearest *thousandth*, the answer must be expressed to *three* decimal digits (three numbers to the right of the decimal point). Look at the fourth decimal digit (fourth number to the right of the decimal point). If the fourth digit is 5 or more, the third decimal digit must be made one digit larger (examples: 0.5982 = 0.598 or 0.5876 = 0.588).

Exercises

Write the correct answers in the spaces provided.

1. Express $\frac{3}{7}$ in decimal notation, rounded to hundredths. _____

2. Carlos, a high school football player, gained 204 yards on 30 carries in a college football game. Find the average number of yards gained per carry. Round to the nearest hundredth.   _____

   a. 6.8

   b. 7.00

   c. 6.80

   d. 5.80

   e. 6.81

3. Earl is 56 years old and is buying $90,000 of life insurance for an annual premium of $693.11. If he pays each annual premium in 12 equal installments, how much is each monthly payment? Round your answer to the nearest cent.   _____

   a. $58.76

   b. $57.75

   c. $57.65

   d. $57.76

   e. $58.86

4. Round 91.642 to the nearest hundredth.   _____

   a. 92.00

   b. 91.64

   c. 91.65

   d. 0.65

   e. 0.64

5. A case of diet cola costs $6.75. If there are 24 cans in a case, find the cost per can. Round to the nearest cent.   _____

   a. $0.38

   b. $0.18

   c. $0.28

   d. $1.48

   e. $1.28

6. Write $\frac{1}{7}$ as a percent (round to the nearest tenth of a percent).   _____

7. What is $16\frac{1}{2}\%$ of 77 (round off answer to hundredths)? _____

8. What is the cost of $\frac{2}{3}$ pound of roast beef at \$3.79 per pound (rounded off to the nearest cent)? _____

9. Change $\frac{7}{12}$ to a decimal, rounded off to hundredths. _____

10. Express $\frac{5}{9}$ as a decimal, rounded to the nearest tenth. _____

11. A television costs \$975. Tien gave a deposit of \$179 and will pay the balance in 12 equal monthly installments. How much is each monthly installment (round your answer to the nearest cent)? _____

12. Round 91.742 to the nearest tenth. _____

13. Represent the product of 3.7×2.6 to the nearest tenth. _____

14. Represent the quotient of 12.2 divided by 0.75 to the nearest hundredth. _____

Chapter 7

Percent Problems

Calculations with Percentages

1. Change the percent to a decimal.

2. What is 12% of 72? Change the percent to a decimal and proceed to multiply; $72 \times 0.12 = 8.64$.

3. If 30% of a number is 60, find the number. Change the percent to a decimal and proceed to divide; $60 \div 0.30 = 200$.

Verbal Problems

Sales Tax and Percent Increase

1. Change the tax (or increase) percent to a decimal.

2. Multiply the original amount by the percent in terms of a decimal. The product is the sales tax (or increase).

3. Add on sales tax (or increase) to the original amount.

$$\text{Total Price} = \text{Original Price} + \text{Sales Tax}$$

Discount and Percent Decrease

1. Change the discount percent to a decimal.

2. Multiply the original price by the percent in terms of a decimal. The product is the discount (or decrease).

3. Subtract the discount (or the decrease) from the original price.

$$\text{Sale Price} = \text{Original Price} - \text{Discount}$$

Write the correct answers in the spaces provided.

1. 9% of what number is 45? _____

2. The circus has 650 employees and hires an additional 12% for the vacation season. What is the total number of employees hired for the vacation season? _____

3. What is 40% of 36.25? _____

4. In this country, 12.3% of the men earn a master's degree whereas only 11.6% of the women earn a master's degree. How many women in this country earn master's degrees? _____

   a. 19,300,000

   b. 12,400,000

   c. 17,000,000

   d. 29,000,000

   e. Not enough information

5. At a family gathering, 3 out of 5 couples disagree on a political matter. What percent of the couples agree on the political matter? _____

   a. 60%

   b. 40%

   c. 20%

   d. 30%

   e. 90%

6. What is 0.6% of 3000?

_____

7. Which is smaller, 80% of 16 or 16% of 85?

_____

8. On a math test, 6 students earned an A. This number is exactly 40% of the total number of students in the class. How many students are in the class?

_____

   a. 15

   b. 24

   c. 19

   d. 30

   e. 14

9. A total of 200 juniors and seniors were given a mathematics test. The 140 juniors attained an average score of 65, and the 60 seniors attained an average score of 75. What was the average score for all 200 students who took the test?

_____

   a. 62

   b. 74

   c. 70

   d. 68

   e. 75

10. The government inspected a meat processing plant and found that on the average, 2 pounds of ground beef out of the 200 pounds inspected did not meet government health code standards. What percent of the ground beef did meet government health code standards?

_____

11. A suit that was originally priced at $450 had its price reduced on Columbus Day by 10%. On Election Day, the price was again reduced by 10%. How much does the suit cost after Election Day? _____

 a. $364.50

 b. $360.00

 c. $540.00

 d. $544.50

 e. $50.00

12. Phil buys a baseball card for $250 and sells it on the Internet for $300. This selling price is an increase of what percent of the original cost? _____

 a. 8%

 b. 13%

 c. 15%

 d. 20%

 e. 24%

13. What is the discount on a coat that sells for $325 if it is reduced by 20% and a month later by 10%? _____

 a. $234.00

 b. $71.50

 c. $91.50

 d. $101.00

 e. $91.00

14. Rent makes up 50% of Avram's household expenses. If his rent is increased by 15% while the rest of his expenses remain the same, by what percentage will his total household expenses increase? _____

 a. 50%

 b. 9%

 c. 8.5%

 d. 7.5%

 e. 5.5%

15. José mistakenly divided the correct sum of his 4 test scores by 5, which gave him a 60 average. What is José's correct average test score? _____

    a. 76

    b. 75

    c. 72

    d. 74

    e. 79

16. 1.54 is what percent of 8? _____

17. Which is larger, 6% of 180 or 75% of 8? _____

18. 45% of what is 13.5? _____

19. What is $12\frac{1}{2}\%$ of 2880? _____

20. Elliot charged $600 on his credit card. He was not charged any interest on his first bill. He made a $60 payment. He then charged an additional $28 for flowers for his wife. On his second bill a month later, he was charged 2% interest on the entire unpaid balance. How much interest was Elliot charged on the second bill? _____

21. What is 375% of 60? _____

22. 196 is $6\frac{1}{4}$% of what? _____

23. Find 12% of 72. _____

24. 12% of what number is 72? _____

25. Fahd left the waitress a 15% tip. If dinner cost $19.00, how much was the tip? _____

26. There is an $8\frac{1}{4}$% sales tax. What is the total cost of a $325 suit? _____

27. A $375 suit is discounted by 20% on Memorial Day. There is an $8\frac{1}{2}$% sales tax. What is the sales tax on the discounted suit? _____

28. Last year, 80% of the graduating class of William Howard Taft High School had taken at least 7 math courses. Of the remaining class members, 75% had taken 5 or 6 math courses. What percent of the graduating class had taken fewer than 5 math courses? _____

    a. 0%

    b. 4%

    c. 5%

    d. 15%

    e. 155%

Chapter 8

Scientific Notation

Scientific notation is the process of expressing a number as the product of two factors. One factor is expressed as a number between 1 and 10. The other factor is a power of 10. The sign of the exponent on the second factor may be positive or negative, as follows:

- Represent 65,439 in scientific notation (Answer: 6.5439×10^4). The decimal point is moved from right to left, and the exponent is positive.

- Represent 0.00065439 in scientific notation (Answer: 6.5439×10^{-4}). The decimal point is moved from left to right, and the exponent is negative.

Exercises

Write the correct answers in the spaces provided.

1. In scientific notation, $75,000 + 4,125,000$ can be represented by _____

    a. 4.2×10^7

    b. 8.4×10^6

    c. 8.4×10^7

    d. 4.88×10^6

    e. 4.2×10^6

2. Write the number 0.05 in scientific notation. _____

    a. 5×10^{-1}

    b. 5×10^{-2}

    c. 5×10^{-3}

    d. 5×10^{-4}

    e. 5×10^{-5}

3. Write the number 0.0069 in scientific notation. _____

    a. 6.9×10^{-1}

    b. 6.9×10^{-2}

    c. 6.9×10^{-3}

    d. 6.9×10^{-4}

    e. 6.9×10^{-5}

4. Write the number 0.0068 in scientific notation. _____

    a. 6.8×10^{-1}

    b. 6.8×10^{-2}

    c. 6.8×10^{-3}

    d. 6.8×10^{-4}

    e. 6.8×10^{-5}

5. Write the number 0.00046 in scientific notation. _____

    a. 4.6×10^{-2}

    b. 4.6×10^{-3}

    c. 4.6×10^{-4}

    d. 4.6×10^{-5}

    e. 4.6×10^{-6}

6. In scientific notation, $60,000 + 3,700,000 = ?$ _____

    a. 2.75×10^{7}

    b. 5.70×10^{6}

    c. 5.70×10^{7}

    d. 3.76×10^{6}

    e. 3.76×10^{7}

7. Write the number 0.0006 in scientific notation. _____

    a. 6×10^{-3}

    b. 6×10^{-4}

    c. 6×10^{-5}

    d. 6×10^{-6}

    e. 6×10^{-7}

8. Write in standard form: 4.27×10^3 _____

9. Write in standard form: 9.34×10^{-2} _____

10. Write in standard form: 1.27×10^4 _____

11. Write in standard form: 6.4×10^{-3} _____

12. Write 7,380,000 in scientific notation. _____

13. Write 0.00759 in scientific notation. _____

14. Write in scientific notation: three million seven hundred thirty-nine thousand. _____

15. Write in scientific notation: three hundred fifty-one thousand. _____

Chapter 9

Statistics

Mean, Median, Mode

Mean (average)   The sum of a set of n numbers divided by n. The symbol for mean is $\bar{x}$, read as x-bar.

Median   Found by arranging a set of data in numerical order from the lowest value to the highest value. The middle value is the median. The number of values that precede the median is equal to the number of values that follow it. If there is an odd number of values (after the data are arranged from smallest to largest), the median is the middle number. For example, find the median of 3, 5, 8, 9, 11, 14, 15. The median is 9. If there is an even number of values, after the data are arranged from smallest to largest, the median is found by adding the two middle terms and dividing by 2. For example, find the median of 3, 4, 7, 9, 11, 13, 16, 17. Add the two middle values and divide by 2 $\left(\dfrac{9 + 11}{2} = \dfrac{20}{2} = 10 \right)$.

Mode   The value that appears the largest number of times in a set of data. List all the values and count the occurrences of each.

Exercises

Write the correct answers in the spaces provided.

1. The number of televisions sold each month for 1 year was recorded by Tommaso's Electronics Store. The results were 16, 13, 25, 25, 22, 19, 26, 27, 19, 25, 16, and 31. Calculate the mean, the median, and the mode of the number of televisions sold per month.

   a. Mean: 22 TVs; median: 23.5 TVs; mode: 25 TVs

   b. Mean: 22 TVs; median: 25 TVs; mode: 23.5 TVs

   c. Mean: 23.5 TVs; median: 23.5 TVs; mode: 23.5 TVs

   d. Mean: 22 TVs; median: 22 TVs; mode: 23.5 TVs

   e. Mean: 23.5 TVs; median: 25 TVs; mode: 22 TVs

2. The numbers of seats occupied on each flight of a jet for 16 trans-Pacific flights were recorded. The numbers were 311, 447, 397, 435, 414, 359, 371, 321, 428, 401, 335, 424, 411, 392, 432, and 410. Calculate the mean, the median, and the mode of the number of seats occupied per flight. _____

   a. Mean: 393 seats; median: 405.5 seats; mode: no mode

   b. Mean: 393 seats; median: 399.25 seats; mode: 405.5 seats

   c. Mean: 405.5 seats; median: 405.5 seats; mode: 399.25 seats

   d. Mean: 399.25 seats; median: 405.5 seats; mode: no mode

   e. Mean: 399.25 seats; median: 405.5 seats; mode: 393 seats

3. The times, in seconds, for a 100-meter dash at a college track meet were 11.06, 10.80, 11.19, 11.62, 10.83, 11.47, 11.31, 11.24, 11.15, and 11.34. Calculate (1) the mean time for the 100-meter dash and (2) the median time for the 100-meter dash. _____

   a. (1) Mean: 11.201 seconds; (2) median: 11.201 seconds

   b. (1) Mean: 11.201 seconds; (2) median: 11.215 seconds

   c. (1) Mean: 11.215 seconds; (2) median: 11.215 seconds

   d. (1) Mean: 11.215 seconds; (2) median: 11.201 seconds

   e. None of the above

4. A consumer research organization purchased identical items in 8 supermarkets. The costs for the purchased items were $44.29, $50.68, $39.84, $39.11, $47.26, $40.83, $46.31, and $44.23. Calculate the mean and the median cost of the purchased items. Round your answers to three decimal places. _____

   a. Mean: $44.260; median: $44.260

   b. Mean: $44.069; median: $44.260

   c. Mean: $44.069; median: $44.069

   d. Mean: $44.260; median: $44.069

   e. None of the above

5. Mary has a 91 average on three mathematics exams. On her first two exams, she received a 94 and a 95. What was her score on the third exam? _____

   a. 84

   b. 93

   c. 91

   d. 94

   e. 89

6. Jacques scored 91, 93, 84, and 87 on the first four of five tests. If Jacques wants an average of exactly 90, what score must he receive on the fifth test?

   a. 88

   b. 95

   c. 93

   d. 97

   e. 100

7. Find the mean of 6, 2, 0, 1, and 6.

8. Find the mode of 7, 2, 3, 5, 7, 5, and 7.

9. Tyrone shoots 4, 5, 3, 4, 3, 2, 3, 5, and 4 on the first nine holes of the U.S. Open. What is his average number of strokes per hole?

10. Find the mean of 0, 0, 60, and 120.

11. During their first year of marriage, Juan bought his wife six pieces of jewelry for a total cost of $1,806. What was the average cost per piece of jewelry?

12. Soraya earned $129 in tips as a waitress working 12 hours. What was the average amount of tips per hour?

13. In the last seven games, a baseball player had the following numbers of hits: 3, 2, 1, 0, 2, 3, and 4. What is the mode?   _____

14. In the last six games, a baseball player had the following numbers of hits: 3, 0, 2, 1, 4, and 1. What is the median?   _____

15. On the highway, a truck driver travels 60 mph for 4 hours and 40 mph for 5 hours. What was the average speed of the truck (round off to the nearest tenth)?   _____

16. A student earned scores of 88, 91, and 82 on the first three of four tests. If the student wants an average of exactly 88, what score must the student earn on the fourth test?   _____

17. A student has grades of 79, 82, and 83 on the first three of four mathematics tests. What must his grade be on the fourth mathematics test to have an 80 average?   _____

   a. 84

   b. 75

   c. 79

   d. 76

   e. 85

Chapter 10

Area, Perimeter, and Cost

Area   The number of square units needed to cover the surface to be measured.

Perimeter   The perimeter of a closed figure, such as a rectangle, is the sum of all the sides.

Formulas for Area

Area of a Rectangle = length × width = (square units).

Area of a Square = length of one side^2 = s^2 = (square units).

Formula for Perimeter

The perimeter of a closed figure is the sum of the lengths of all its sides.

Perimeter of a Rectangle = 2 lengths + 2 widths

Write the correct answers in the spaces provided.

1. Find the area of a room that is 8 yards by 5 yards.   _____

2. Find the area of a tablecloth that is 65 inches long and 30 inches wide.   _____

3. How much does it cost to carpet a room 9 yards by 8 yards at $7.00 per square yard?

_____

4. Carpeting costs $9.00 per square yard. How much does it cost to carpet a room 7 yards by 5 yards?

_____

5. Find the perimeter of a rectangle whose length is 9 feet and whose width is 6 feet.

_____

6. How much fencing is needed to fence in a playground 35 feet long and 20 feet wide?

_____

7. Fencing costs $6.00 per yard. How much does it cost to fence in a rectangular yard that is 40 yards by 30 yards?

_____

8. How much does it cost to fence in a rectangular garden 18 feet by 9 feet at $7.00 per foot of fencing?

_____

9. Carpeting costs $12.00 per square yard. How much does it cost to carpet a room 9 yards by 8 yards?

_____

10. Find the perimeter of a triangle whose sides are 3, 4, and 5 inches. _____

11. Find the perimeter of a square whose side measures 1.9 meters. _____

    a. 3.8 meters

    b. 5.7 meters

    c. 7.6 meters

    d. 3.61 meters

12. Phyllis is planting tomato seeds in her garden. The garden's dimensions are 12 yards by 9 yards. Each bag of seeds covers 11 square yards. How many bags of seeds should she buy? _____

    a. 10

    b. 9

    c. 11

    d. 8

    e. None of these

13. Hoshi is adding topsoil to her flower garden, which measures 20 yards by 18 yards. Each bag covers 14 square yards and costs $11.95. What is the total cost of the topsoil for the flower garden? _____

    a. $298.75

    b. $310.70

    c. $299.55

    d. $310.50

    e. None of these

Chapter 11

Ratios and Proportions

Key Terms

Proportion A statement that two ratios (or fractions) are equal, based on the Means-Extremes Property, as follows:

If a, b, c, and d are real numbers, when $b \neq 0$ and $d \neq 0$, then

if $\dfrac{a}{b} = \dfrac{c}{d}$

then $ad = bc$.

In other words, in any proportion, the product of the means ($a \times d$) is equal to the product of the extremes ($b \times c$).

Ratio If x and y are any two numbers, where $y \neq 0$, then the ratio of x and y is $\dfrac{x}{y}$.

Unit rate A ratio (or rate) in which the second (bottom) term equals 1.

Exercises

Write the correct answers in the spaces provided.

1. Find the value of x that solves the proportion: $\dfrac{9}{4} = \dfrac{11}{x}$ _____

   a. 5

   b. $5\dfrac{1}{9}$

   c. $3\dfrac{1}{4}$

   d. $4\dfrac{8}{9}$

   e. $5\dfrac{3}{4}$

2. If the total cost of a apples is c cents, what is the general formula for the cost, in cents, of b apples? _____

    a. $\dfrac{ab}{c}$

    b. $\dfrac{a}{bc}$

    c. $\dfrac{c}{ab}$

    d. $\dfrac{ac}{b}$

    e. $\dfrac{bc}{a}$

3. Solve. Round to the nearest hundredth, if necessary. _____

$$\frac{3}{20} = \frac{x}{18}$$

    a. 2.70

    b. 2.75

    c. 3.75

    d. 3.72

    e. None of the above

4. Write the phrase as a rate in simplest form. _____

402 feet in 24 seconds

    a. $\dfrac{67 \text{ ft}}{4 \text{ sec}}$

    b. $\dfrac{67 \text{ sec}}{4 \text{ ft}}$

    c. $\dfrac{268 \text{ ft}}{16 \text{ sec}}$

    d. $\dfrac{268 \text{ sec}}{16 \text{ ft}}$

    e. $\dfrac{134 \text{ sec}}{8 \text{ ft}}$

5. Determine whether the proportion is true or false. _____

$$\frac{84}{60} = \frac{30}{18}$$

    a. True

    b. False

6. Solve: $\dfrac{8}{x} = \dfrac{4}{7}$

   a. 42

   b. 3.5

   c. 14

   d. 28

   e. None of the above

7. Determine whether the proportion is true or false.

$$\frac{15 \text{ ft}}{50 \text{ sec}} = \frac{15 \text{ ft}}{120 \text{ sec}}$$

   a. True

   b. False

8. Solve: $\dfrac{72}{84} = \dfrac{30}{x}$

   a. 3.5

   b. 35

   c. 21

   d. 42

   e. None of the above

9. Arturo Rodriguez owns 70 shares of Texas Utilities that pay dividends of $130. At this rate, what dividend would Arturo receive after buying 560 additional shares of Texas Utilities?

   a. $1070

   b. $1320

   c. $970

   d. $1120

   e. None of the above

10. A car travels 75.69 miles on 3 gallons of gas. Find the distance that the car can travel on 11 gallons of gas. Round to the nearest hundredth.

   a. 277.35 miles

   b. 277.53 miles

   c. 27.83 miles

   d. 2.78 miles

   e. None of the above

11. Allison Yee bought a computer system for $7200. Five years later, she sold the computer for $1600. Find the ratio of the amount she received for the computer to the cost of the computer.

          _____

  a. $\dfrac{2}{9}$

  b. $\dfrac{9}{2}$

  c. $\dfrac{4}{9}$

  d. $\dfrac{9}{4}$

  e. $\dfrac{2}{18}$

12. The average price of a 60-second commercial during a popular sitcom last year was $814,800. Find the price per second.

          _____

  a. $\dfrac{13{,}580 \text{ sec}}{\$1}$

  b. $\dfrac{\$13{,}580}{\text{sec}}$

  c. $\dfrac{\$407{,}400}{30 \text{ sec}}$

  d. $\dfrac{407{,}400 \text{ sec}}{\$30}$

  e. $\dfrac{\$814{,}800}{60 \text{ sec}}$

13. The price of gasoline jumped from $1.12 to $1.44 in 1 year. What is (1) the increase in the price of gasoline and (2) the ratio of the increase in price to the original price?

          _____

  a. (1) $0.32   (2) $\dfrac{7}{2}$

  b. (1) $2.56   (2) $\dfrac{2}{7}$

  c. (1) $0.32   (2) $\dfrac{2}{7}$

  d. (1) $2.56   (2) $\dfrac{7}{2}$

  e. (1) $0.64   (2) $\dfrac{1}{7}$

14. A bank requires a monthly payment of $31.85 on a $2900 loan. At the same rate, find the monthly payment on an $8700 loan. Round to the nearest hundredth.

    a. $95.55

    b. $90.05

    c. $90.55

    d. $105.55

    e. None of the above

15. Write the comparison as a ratio in simplest form using a fraction, a colon (:), and the word *to*.

132 dollars to 108 dollars

    a. $\dfrac{11}{9}$ 11:9   11 to 9

    b. $\dfrac{9}{11}$ 9:11   9 to 11

    c. $\dfrac{44}{36}$ 44:36   44 to 36

    d. $\dfrac{36}{44}$ 36:44   36 to 44

    e. $\dfrac{33}{27}$ 33:27   33 to 27

16. An automobile was driven 359.52 miles on 11.2 gallons of gas. Find the number of miles driven per gallon of gas.

    a. $\dfrac{32.1 \text{ gal}}{\text{mi}}$

    b. $\dfrac{64.2 \text{ mi}}{\text{gal}}$

    c. $\dfrac{64.2 \text{ gal}}{\text{mi}}$

    d. $\dfrac{32.1 \text{ mi}}{\text{gal}}$

    e. None of these

17. Write the phrase as a rate in simplest form. _____

$$756 \text{ for } 48 \text{ packs}$$

a. $\dfrac{\$756}{48}$

b. $\dfrac{\$4}{63 \text{ packs}}$

c. $\dfrac{\$63}{4 \text{ packs}}$

d. $\dfrac{\$16}{252 \text{ packs}}$

e. $\dfrac{\$315}{20 \text{ packs}}$

18. Write the phrase as a rate in simplest form. _____

385 gallons in 19 hours

a. $\dfrac{1155 \text{ hr}}{57 \text{ gal}}$

b. $\dfrac{385 \text{ hr}}{19 \text{ gal}}$

c. $\dfrac{1155 \text{ gal}}{57 \text{ hr}}$

d. $\dfrac{385 \text{ gal}}{19 \text{ hr}}$

e. $\dfrac{2310 \text{ hr}}{114 \text{ gal}}$

19. A house with an original value of $100,000 increased in value to $140,000 in 5 years. Find (1) the increase in the value of the house and (2) the ratio of the increase in value to the original value of the house. _____

a. (1) $240,000   (2) $\dfrac{2}{5}$

b. (1) $40,000   (2) $\dfrac{2}{5}$

c. (1) $40,000   (2) $\dfrac{5}{2}$

d. (1) $240,000   (2) $\dfrac{5}{2}$

e. (1) $20,000   (2) $\dfrac{1}{5}$

20. Solve. Round to the nearest hundredth, if necessary.

$$\frac{x}{18} = \frac{6}{4}$$

    a. 27.05

    b. 37.05

    c. 25.95

    d. 27.00

    e. None of the above

21. Write the phrase as a unit rate: 168 feet in 12 seconds.

    a. $\dfrac{14 \text{ ft}}{\text{sec}}$

    b. $\dfrac{14 \text{ sec}}{\text{ft}}$

    c. $\dfrac{84 \text{ ft}}{6 \text{ sec}}$

    d. $\dfrac{84 \text{ sec}}{6 \text{ ft}}$

    e. $\dfrac{112 \text{ sec}}{8 \text{ ft}}$

22. You own 360 shares of stock in a computer company. The company declares a stock split of 5 shares for every 2 owned. How many shares of stock will you own after the stock split?

    a. 720

    b. 144

    c. 900

    d. 441

    e. None of the above

23. Solve. Round to the nearest hundredth, if necessary.

$$\frac{33}{x} = \frac{2}{3}$$

    a. 49.60

    b. 49.70

    c. 49.37

    d. 51.37

    e. None of the above

24. You drove 393.30 miles in 9 hours. Find the average number of miles you drove per hour.

a. $\dfrac{43.7 \text{ mi}}{\text{hr}}$

b. $\dfrac{43.7 \text{ hr}}{\text{mi}}$

c. $\dfrac{196.65 \text{ mi}}{4.5 \text{ mi}}$

d. $\dfrac{196.65 \text{ hr}}{4.5 \text{ mi}}$

e. $\dfrac{485.07 \text{ hr}}{11.1 \text{ mi}}$

25. Ron Stokes uses 2 pounds of fertilizer for every 200 square feet of lawn for landscape maintenance. At this rate, how many pounds of fertilizer did he use on a lawn that measures 900 square feet?

a. 45

b. 60

c. 75

d. 90

e. None of these

26. Write the phrase as a unit rate.

$146.40 earned in 8 minutes

a. $\dfrac{\$18.30}{\text{min}}$

b. $\dfrac{18.3 \text{ min}}{\$1}$

c. $\dfrac{\$146.40 \text{ min}}{8 \text{ min}}$

d. $\dfrac{146.40 \text{ min}}{\$8}$

e. $\dfrac{109.80 \text{ min}}{\$6}$

27. Determine whether the proportion is true or false.

$$\frac{5 \text{ rolls}}{6 \text{ ft}} = \frac{10 \text{ rolls}}{12 \text{ ft}}$$

    a. True

    b. False

28. On a map, two towns are $2\frac{1}{4}$ inches apart. If $\frac{3}{8}$ inch on the map represents 30 miles, find the number of miles between the two towns.

29. Two ounces of vanilla ice cream contains 39 calories. How many calories are there in 11 ounces of vanilla ice cream?

    a. 195

    b. 214.5

    c. 238

    d. 114.5

    e. 107.5

30. Solve: $\dfrac{x}{4} = \dfrac{-2}{3}$

PART II

ALGEBRA AND GEOMETRY

Chapter 12

Rules for Signed Numbers

Key Terms

Base   The number that serves as a starting point:

Base^Power

3^4:   3 base, 4 power

6^7:   6 base, 7 power

Difference   Means subtraction.

Exponent   The number of times a factor is used.

Power   The number of times a base is used as a factor.

Product   Means multiplication.

Quotient   Means division.

Squaring a Number   A number multiplied by itself.

Sum   Means addition.

Variable   A symbol that stands for a number in an algebraic expression.

Addition

When adding numbers with the same signs, keep the sign and add the numbers.

When adding numbers with opposite signs, keep the sign of the larger number and subtract the numbers.

Subtraction

When subtracting numbers, change the sign of the subtrahend (the number after the subtraction sign) to the opposite sign and then follow the rules for addition.

Multiplication and Division

When multiplying or dividing *two* numbers with the *same* signs, the result is positive.

When multiplying or dividing *two* numbers with *opposite* signs, the result is negative.

The product of an odd number of negative numbers is negative.

The product of an even number of negative numbers is positive.

Division Facts

$\dfrac{a}{a} = 1$ A non-zero number divided by itself equals 1.

$\dfrac{a}{1} = a$ A number divided by 1 equals the number.

$\dfrac{0}{a} = 0$ Zero divided by a non-zero number equals zero.

$\dfrac{a}{0} = \varnothing$ Undefined. A number divided by zero is undefined—not defined.

Order of Operations

1. Perform indicated operations inside the parentheses.
2. Evaluate the roots and the powers.
3. Multiply or divide from left to right.
4. If necessary, add or subtract.

Exercises

Write the correct answers in the spaces provided.

1. During the day, the temperature increased from –12°C to 24°C. What was the increase in temperature?   _____

2. Tamara was three floors below the lobby. The elevator was broken so she walked up the stairs to the 12th floor. How many floors did she walk?   _____

3. During the week, Michael's investments in the stock market fluctuated as follows: On Monday, he lost \$300; on Tuesday, he gained \$600; on Wednesday, he gained \$100; on Thursday, he gained \$700; and on Friday, he lost \$400. What was the total amount of money Michael lost or gained? _____

4. During the last four weeks, Nikolai was on the Fitness Diet and lost 3 pounds the first week, $1\frac{1}{4}$ pounds the second week, 2 pounds the third week, and $1\frac{1}{2}$ pounds the fourth week. If he originally weighed 205 pounds, how much does he weigh after the diet? _____

Evaluate the expressions in Exercises 5–19:

5. $2 - 3(7 - 4)$ _____

6. $6 + 2(8 + 4)$ _____

7. $3(8 + 1) + 4$ _____

8. $7(4 + 3) - 3(5 + 9)$ _____

9. $4(6)^2$ _____

10. $-3(4 + 5)^2$ _____

11. $-3 + (4 + 5)^2$ _____

12. $6 - 7(5)^2$ _____

13. $-3(-4)^2$ _____

14. $5^2 + 3^0$ _____

15. $7(6 + 5)^2$ _____

16. $3 + 2(4)^3$ _____

17. $(3 + 4)^0$ _____

18. $7 + 2(6 + 5)^2$ _____

19. $6 + 5(7)^3$ _____

20. Normal systolic blood pressure is sometimes approximated by using a person's age, in years, plus 100. Approximate the difference between the systolic blood pressure of a 12-year-old and the systolic blood pressure of that same person at the age of 60. _____

    a. 272

    b. 96

    c. 38

    d. 28

    e. 48

21. Seth was 9 floors below the lobby of the New York Building. He then walked up through the lobby, beyond the first floor and all the way up to the 12th floor. How many floors did Seth climb? _____

    a. 26 floors

    b. 21 floors

    c. 20 floors

    d. 29 floors

    e. 25 floors

22. In Detroit, there are 180 days in a school year. Zelmita was absent N days. How many days was she present? _____

23. A computer technician charges \$75 per hour. If he worked h hours, how much did he get paid? _____

24. If Mr. Dinesh travels m miles in x hours, how many miles does he travel per hour? _____

25. Assume

One dime = 10 cents; "d" dimes = 10d

One quarter = 25 cents; "q" quarters = 25q

Express d dimes and q quarters in cents.

26. Each week, Jamila gets paid R dollars and a 5% commission on her total sales S. What is Jamila's weekly pay?

a. $R + S$

b. $R + 0.05S$

c. $0.05R + 0.05S$

d. $S + 0.05R$

e. $R - S$

27. Translate this phrase into a mathematical expression: The product of x and the sum of x and 5.

28. Express algebraically the following: the square of a number plus the product of 6 and the number.

29. Translate the following into a mathematical expression: the quotient of x and the difference between x and 12.

30. Assume

One nickel = 5 cents; "n" nickels = 5n

One quarter = 25 cents; "q" quarters = 25q

Express n nickels and q quarters as an algebraic expression.

Chapter 13

Exponent Rules

For Multiplication

When multiplying numbers with the same base, keep the base and add the exponents:

$$(A^m)(A^n) = A^{m+n}$$

If no exponent is written, it is understood to be 1.

For Division

When dividing numbers with the same base, keep the base and subtract the exponents (exponent in the numerator minus exponent in the denominator).

$$\frac{x^m}{x^n} = x^{m-n}$$

For Raising a Number to a Power

To raise a number (x^m) to a power n, multiply the exponents.

$$(x^m)^n = x^{mn}$$

For Eliminating Negative Exponents

If the negative exponent is in the numerator, bring it to the denominator as a positive exponent.

$$x^{-m} = \frac{1}{x^m} \qquad x^{-3} = \frac{1}{x^3}$$

If the negative exponent is in the denominator, bring it to the numerator as a positive exponent.

$$\frac{1}{x^{-m}} = x^m \qquad \frac{1}{x^{-4}} = x^4$$

Exercises

Write the correct answers in the spaces provided.

Simplify your answers.

1. $(x^2)^3(x) =$ _____

2. $(x^2)(y^3)(x^3)(y^2) =$ _____

3. $(x^{-2})^3 =$ _____

4. $\dfrac{x^7}{x^9} =$ _____

5. $\dfrac{x^9}{x^7} =$ _____

6. $(x^{-3}y^4)^2 =$ _____

7. $\dfrac{x^{-2}}{y^{-3}} =$ _____

8. $\dfrac{y^{-7}}{y^{-3}} =$ _____

9. $x^0 =$ _____

10. $\dfrac{3x^{-2}}{(2x)^{-3}} =$ _____

11. Simplify: $(-2x^6)(3x^{-5})^2$ _____

12. Simplify: $(-3y^3)(2y^{-7})^2$ _____

    a. $\dfrac{-6}{y^{17}}$

    b. $-36y^{11}$

    c. $\dfrac{-36}{y^{11}}$

    d. $\dfrac{12}{y^{11}}$

    e. $\dfrac{-12}{y^{11}}$

13. For all $x \neq 0$ and $y \neq 0$, $\dfrac{x^{-3}y^2}{x^5y^{-4}} = ?$ _____

    a. $\dfrac{x^2}{y^2}$

    b. $\dfrac{y^3}{x^4}$

    c. $\dfrac{y^6}{x^2}$

    d. $\dfrac{y^6}{x^8}$

    e. $\dfrac{1}{x^2y}$

14. For all x, y, and z, $(-3x^3y^2z)^2 = ?$ _____

    a. $-6x^5y^4z^2$

    b. $-9x^6y^4z^2$

    c. $9x^6y^4z^2$

    d. $6x^5y^4z^3$

    e. $9xyz$

15. For all $x \neq 0$ and $y \neq 0$, $\dfrac{x^{-4}y^3}{x^6y^{-5}} = ?$ _____

    a. $\dfrac{y^8}{x^{10}}$

    b. x^2y^2

    c. x^2y^8

    d. x^8y^2

    e. $\dfrac{x^{10}}{y^8}$

16. Simplify: $\dfrac{3x^{-3}}{(2x)^{-4}}$ _____

17. Simplify: $\dfrac{2y^{-2}}{(2y)^{-3}}$ _____

18. Simplify: $\dfrac{7x^{-4}B^{-3}}{17x^{12}B^{14}}$ _____

19. Simplify: $\dfrac{-6A^{-3}B^{-7}}{3A^{-4}B^{-6}}$ _____

    a. $\dfrac{-A}{2B}$

    b. $-2AB$

    c. $\dfrac{-2B}{A}$

    d. $\dfrac{-2A}{B}$

    e. $\dfrac{-B}{2A}$

20. Evaluate: $16^{\frac{-3}{4}}$ _____

    a. $\frac{1}{8}$

    b. 8

    c. 64

    d. $\frac{1}{64}$

    e. $\frac{1}{16}$

21. Evaluate: 2^5 _____

    a. 10

    b. 32

    c. 4

    d. 8

    e. 16

22. Evaluate: $(-2)^3(-4)^2$ _____

    a. −32

    b. −128

    c. 128

    d. −256

    e. 32

23. Evaluate: $2(-6)^3$ _____

    a. −72

    b. 72

    c. −432

    d. 432

    e. 234

24. Simplify: $49^{3/2}$ _____

    a. 7

    b. 49

    c. 343

    d. $\dfrac{1}{343}$

    e. $\dfrac{1}{49}$

25. Simplify: $(-36)^{1/2}$ _____

    a. −6

    b. 6

    c. $-\dfrac{1}{6}$

    d. $\dfrac{1}{6}$

    e. Not a real number

26. Simplify: $36^{3/2}$ _____

    a. 6

    b. 216

    c. 36

    d. $\dfrac{1}{6}$

    e. $\dfrac{1}{36}$

27. Simplify: $(9)^{3/2}$ _____

    a. −27

    b. 27

    c. $-\dfrac{1}{27}$

    d. $\dfrac{1}{27}$

    e. Not a real number

28. Simplify: $27^{1/3}$ _____

    a. 2

    b. 8

    c. 4

    d. $\dfrac{1}{3}$

    e. 3

29. Simplify: $216^{2/3}$ _____

    a. 36

    b. 216

    c. -36

    d. $\dfrac{1}{6}$

    e. $\dfrac{1}{36}$

30. Simplify: $125^{1/3}$ _____

    a. 5

    b. 25

    c. 125

    d. $\dfrac{1}{5}$

    e. $\dfrac{1}{25}$

31. Simplify: $8^{-2/3}$ _____

    a. 2

    b. 8

    c. -4

    d. $\dfrac{1}{2}$

    e. $\dfrac{1}{4}$

32. Simplify: $16^{3/2}$ _____

    a. 4

    b. 16

    c. 64

    d. $\dfrac{1}{64}$

    e. $\dfrac{1}{16}$

33. Evaluate: $2(-2)^4$ _____

    a. −32

    b. 32

    c. −64

    d. 64

    e. 16

34. Evaluate: 4^3 _____

    a. 12

    b. 256

    c. 48

    d. $\dfrac{1}{64}$

    e. 64

35. Evaluate: $(-3)^4$ _____

    a. −81

    b. 12

    c. −12

    d. 243

    e. 81

36. Simplify: $(-9)^{1/2}$

    a. -3

    b. 2

    c. $-\dfrac{1}{3}$

    d. $\dfrac{1}{3}$

    e. Not a real number

37. Simplify: $\left(\dfrac{25}{49}\right)^{-3/2}$

    a. $\dfrac{125}{343}$

    b. $\dfrac{343}{125}$

    c. $-\dfrac{125}{343}$

    d. $-\dfrac{343}{125}$

    e. Not a real number

38. Evaluate: $(-6)^3$

    a. $\dfrac{-2}{6}$

    b. $\dfrac{2}{6}$

    c. 36

    d. -36

    e. -216

39. Evaluate: -2^4

    a. 8

    b. -8

    c. 16

    d. -16

    e. -32

Chapter 14

Operations with Polynomials

Key Terms

Binomial   A polynomial consisting of two terms. Example: $4x^2 - 4$.

Monomial   A polynomial consisting of one term. Example: $3x^2$.

Polynomial   A mathematical expression of one or more terms with constants, variables, and exponents.

Addition and Subtraction of Algebraic Expressions

Add or subtract coefficients (numbers in front) of like terms.

Like terms are:

- Same variables: $ax + bx = (a + b)x$

OR

- Same exponents for the same variables: $ax^2 + bx^2 = (a + b)x^2$

Multiplication of a Monomial by a Monomial

1. Determine the sign of the product.

2. Multiply all the coefficients (the numbers in front of the unknowns).

3. Multiply the variables by adding exponents, if the bases are the same. (If no exponent is written, it is understood to be 1.)

Example: $(7x^2y)(-4xy^2)$

Answer: $-28x^3y^3$

Multiplication of a Monomial by a Binomial

Multiply the monomial by each term of the binomial.

Use the Distributive Law of Multiplication, which mandates multiplication before addition. That is, for any three numbers a, b, and c, perform multiplication before addition: $a(b + c) = ab + ac$.

Example: $-3(x + 4)$

Answer: $-3x - 12$

Multiplication of a Binomial by a Binomial

Multiply each term in the first binomial with each term in the second binomial and add all of the resulting products.

Example: $(x - 4)(x - 7)$

Answer: $x^2 - 4x - 7x + 28$

$\quad\quad\quad x^2 - 11x + 28$

Multiplication of Polynomials

To find the product of two polynomials, use the column method—the same method used in long multiplication with whole numbers.

Order of Operations in Algebra
To simplify, first perform multiplication and then addition and subtraction.

Division of a Monomial by a Monomial

1. Divide each part of the numerator by the denominator.

2. Determine the sign of the quotient.

3. Divide the coefficients.

4. Divide the variables by subtracting the exponents, if the bases are the same.

Example: $\dfrac{-4x^7}{2x}$

Answer: $-2x^6$

Hint For the same base in division, if the larger exponent is in the numerator, the answer goes to the numerator with a positive exponent. If the larger exponent is in the denominator, the answer goes to the denominator with a positive exponent.

Exercises

Write the correct answers in the spaces provided.

1. What is the sum of the polynomials $4x^3y - 3x^2y^3$ and $-7x^2y^3 - 7x^3y$?

2. What is the sum of the polynomials $-7x^2 + 3x$ and $9x^3 - 6$?

3. Simplify: $(x - y^2) - (-3xy + y^2)$

4. Simplify: $(x^2 - 7) - (-x^2 + 7)$

5. Translate the following mathematical expression: six more than three times a number and the sum of eight and a number.

6. Translate the following mathematical expression: subtract three more than twice a number from six less than the number.

7. Simplify: $(-7rst)(-2rst)$

8. Simplify: $(-3x^2y^4)^3$ _____

9. Simplify: $(3x^2y)(-4x^5y^4)^3$ _____

10. Simplify: $(9xy^2)(3x^2y^{-4})^3$ _____

11. Simplify: $-7x - x(x - 7)$ _____

12. Simplify: $-7a^2b - 4b(3ab - 6a^2)$ _____

    a. $-9a^2b - 8ab^2$

    b. $8ab^2 + 9a^2b$

    c. $17a^2b + 12ab^2$

    d. $-12ab^2 - 17a^2b^2$

    e. $-12ab^2 + 17a^2b$

13. Simplify: $(2x - 7)(x - 3)$ _____

14. Simplify: $(-x + 7)^2$ _____

15. Simplify: $(x + y)^2 - (9xy - 6x^2)$ _____

    a. $7x^2 + 7xy + y^2$

    b. $7x^2 + y^2$

    c. $7x^2 - 7xy + y^2$

    d. $7x^2 + 11xy + y^2$

    e. $7x^2 - 11xy + y^2$

16. Simplify: $(x + y)^2 - (6xy - 9x^2)$ _____

    a. $9x^2 - 4xy + y^2$

    b. $10x^2 + 4xy - y^2$

    c. $9x^2 + 8xy + y^2$

    d. $10x^2 - 8xy + 9y^2$

    e. $10x^2 - 4xy + y^2$

17. Simplify: $(3 + x)(x - 3)^2$ _____

18. Reduce to lowest terms: $\dfrac{15x^5 - 10x^3 + 5x^2}{5x^2}$ _____

    a. $3x^7 - 2x^5 + x^4$

    b. $3x^3 - 2x$

    c. $2x^5 - x^4 + 3x^3$

    d. $3x^3 + 2x$

    e. $3x^3 - 2x + 1$

19. Divide: $\dfrac{9x^7 + 3x^5 + 3x^2}{3x^2}$ _____

20. Represent the product of four less than twice a number and six more than four times the number. _____

21. For all nonzero x and y, $\dfrac{-6x^7 y^2}{8x^2 y^3} = ?$ _____

    a. $\dfrac{-6x^5}{8y}$

    b. $\dfrac{-3x^5}{4y}$

    c. $\dfrac{3x^5}{4y}$

    d. $\dfrac{-3x^9 y^5}{4}$

    e. $\dfrac{-3x}{4y}$

22. Which of the following expressions represents the product of six more than two times x and four less than three times x?

    a. $6x^2 + 10x - 24$

    b. $6x^2 - 10x - 24$

    c. $6x^2 + 10x + 24$

    d. $6x^2 - 10x + 24$

    e. None of the above

23. Write an expression that represents the quotient of two less than a number y divided by three more than twice the number y. _____

24. Write an expression that represents the quotient of three less than a number x divided by four more than the number x. _____

Chapter 15

Evaluating Algebraic Expressions and Formulas

Order of Operations

1. Substitute the value for the unknown in the given expression or formula. Put the value in parentheses.

2. Evaluate the roots and powers.

3. Multiply or divide from left to right.

4. If necessary, add or subtract.

Exercises

Write the correct answers in the spaces provided.

1. Evaluate $4x^2 - 7xy$ when $x = -4$ and $y = -3$. _____

2. Evaluate $2y^3 - 4xy$ when $x = 2$ and $y = -1$. _____

3. If $x = 4$ and $y = -2$, evaluate $x^3y - y^3$. _____

   a. 120

   b. −136

   c. −120

   d. 0

   e. −16

4. If $x = -2$ and $y = -3$, what is the value of the expression $(x - y)^3$? _____

5. If $x = 2$ and $y = 4$, find the value of $\dfrac{xy - y}{2x}$. _____

   a. 4

   b. −1

   c. −4

   d. 2

   e. 1

6. If $x = -3$ and $y = -1$, what is the value of the expression $-y^3 x$? _____

7. If $x = -2$ and $y = -4$, what is the value of $\dfrac{x^2 y - x}{3x}$? _____

8. Evaluate $\sqrt{x^2 - y^2}$ when $x = 15$ and $y = 9$. _____

9. Evaluate $\sqrt{x^2 - y^2}$ when $x = 10$ and $y = 6$. _____

10. Evaluate the expression: $A = p(1 + r)$. If $p = \$650$ and $r = 9\%$, $A = ?$ _____

   a. \$608.50

   b. \$708.50

   c. \$58.50

   d. \$585.50

   e. \$698.50

11. Evaluate $(3x)^0$ when $x = 4$. _____

   a. −12

   b. 12

   c. −7

   d. 1

   e. −1

12. Evaluate the expression: $P = 2l + 2w$. If $P = 200$ feet and $l = 60$ feet,
$w = ?$

_____

13. Evaluate the expression: $I = Prt$. If $P = \$400$, $r = 8\%$, and $t = 1$ year,
$I = ?$

_____

14. Evaluate the expression: $A = s^2$. If $A = 144$ square feet, $s = ?$

_____

15. Evaluate the expression: $A = s^2$. If $s = 6$ feet, $A = ?$

_____

Chapter 16

Solving Linear Equations

Key Terms

Addition Property   If the same quantity is added to both sides of an equation, the sides remain equal. The Additive Inverse: For every number A, there is exactly one number $-A$ such that $A + (-A) = 0$.

Equation   A statement of equality, consisting of three parts: the right side, the left side, and the equal sign.

Literal Equation   An equation with two or more variables.

Multiplication Property   If both sides of an equation are multiplied by the same quantity, the sides remain equal. The Multiplicative Inverse is the number or expression by which a number or expression is multiplied to equal 1:

$$n \times \frac{1}{n} = 1$$

Solving Linear Equations

1. To solve an equation, use the Addition Property, as follows:

$$X - 2 = 3$$

   Add 2 (the additive inverse of –2) to both sides.

$$X = 5$$

2. If necessary, use the Multiplication Property, as follows:

$$5x = 45$$

   Multiply both sides by 1/5 (the multiplicative inverse of 5).

$$x = 9$$

Solving Equations with Fractions

1. Multiply each and every number and variable on both sides of the equation by the lowest common denominator (LCD).

2. Now follow the above rules used for solving linear equations.

Solving Literal Equations

1. Isolate the variable to be solved for, using the Addition and Multiplication Properties above.

2. Follow the rules used to solve linear equations.

Exercises

Write the correct answers in the spaces provided.

1. Solve: $3x + 6 + 2x = 31$ _____

2. Solve: $-3(x + 4) = -2(x - 17)$ _____

3. Solve: $-4(2x - 7) + 7x = 6$ _____

4. Solve: $-2x + 7(x + 4) = 28$ _____

5. If $-4(x + 2) = -3x - 10$, then $x = ?$ _____

   a. 2

   b. -2

   c. $-\dfrac{1}{2}$

   d. 4

   e. $\dfrac{1}{2}$

6. Solve: $2x + 1.3x = -0.7x + 32$ _____

7. Solve: $3x + 2(x + 12) = -3(x + 8)$
_____

8. Solve: $5 - 4(y - 1) = 26(y - 1)$
_____

9. If $-5 + 2x = -(x - 3) + 2$, then $x = ?$
_____

   a. $3\frac{1}{3}$

   b. 4

   c. $2\frac{1}{3}$

   d. $-3\frac{1}{3}$

   e. 3

10. Solve: $y + 7(y + 4) = 28$
_____

11. Solve: $\dfrac{3}{x} + \dfrac{4}{x} = 20$
_____

12. Solve: $\dfrac{4}{7}y + 3 = 1$
_____

13. Solve: $17(y - 5) + 2y = -8 - 2(y - 5)$
_____

14. Solve: $3(2y + 5) - 4(y - 2) = 3(2 + 2y) + 1$ _____

   a. −4

   b. 4

   c. −3

   d. 3

   e. 6

15. Solve: $\dfrac{1}{x} + \dfrac{4}{x} = 40$ _____

16. Solve: $x - \dfrac{9}{10} = \dfrac{-1}{10}$ _____

17. Solve: $\dfrac{4x}{3} = \dfrac{x - 14}{6}$ _____

18. Solve: $\dfrac{x}{2} + \dfrac{x}{3} = 5$ _____

19. Solve: $\dfrac{x}{3} - 1 = \dfrac{x}{4}$ _____

20. Solve: $\dfrac{5y}{3} = y - 2$ _____

   a. 3

   b. −3

   c. 6

   d. −6

   e. 8

21. Solve for y: $\dfrac{3y}{4} + \dfrac{1}{5} = \dfrac{y}{2}$ _____

   a. 4

   b. $-\dfrac{1}{15}$

   c. $\dfrac{4}{15}$

   d. $\dfrac{4}{5}$

   e. $-\dfrac{4}{5}$

22. Solve for x: $\dfrac{3x}{2} + \dfrac{2}{5} = \dfrac{x}{2}$ _____

   a. $\dfrac{2}{5}$

   b. $\dfrac{-2}{5}$

   c. $\dfrac{4}{15}$

   d. $\dfrac{-1}{15}$

   e. 4

23. If $ax + by + cd = e$, then $y = ?$ _____

24. If $c + by = d$, then $y = ?$ _____

   a. $\dfrac{d + c}{d}$

   b. $\dfrac{d - b}{c}$

   c. $\dfrac{d + b}{c}$

   d. $\dfrac{c - d}{b}$

   e. $\dfrac{d - c}{b}$

25. Solve for r: $F = \dfrac{mv^2}{gr}$ _____

   a. $r = \dfrac{Fg}{mv^2}$

   b. $r = \dfrac{mv^2 Fg}{Fg}$

   c. $r = \dfrac{gmv^2}{F}$

   d. $r = \dfrac{F}{gmv^2}$

   e. $r = \dfrac{mv^2}{Fg}$

26. Solve for l: $P = 2l + 2w$ _____

27. Solve for r: $A = \pi r^2$ _____

28. Solve for x: $ax^2 - bc = d$ _____

Chapter 17

Solving Verbal Problems

Translate Words into Mathematical Expressions

English	*Algebra*
The sum of x and y	$x + y$
The product of x and y	xy
The quotient of x and y	$\dfrac{x}{y}$
A number	x
6 more than a number	$6 + x$
4 times a number	$4x$
The difference of x and y	$x - y$

Exercises

Write the correct answers in the spaces provided.

1. Five times the sum of a number and 6 is 40. Find the number. _____

2. One number is 6 less than another. Their sum is 12. Find both numbers. _____

3. One number is 10 less than another. Their sum is 50. Find the smaller number. _____

4. It costs a manufacturer $20 to produce a video game. In addition, the manufacturer has an overhead of $30,000. The manufacturer produces 6000 games and sells each game for $50. How many games must the manufacturer sell to break even?

_____

5. If $\dfrac{1}{3}$ of a number is 6, what is the value of $6\dfrac{1}{3}$ times the number?

_____

6. This year, the value of a summer cottage is $135,000. This amount is 3 times the value of the cottage 8 years ago. What was the value 8 years ago?

_____

7. A store manager marks up all clothing 30%. What did the store pay for a blouse that sells for $66.95 (round your answer to the nearest cent)?

_____

8. An exit poll survey shows that 3 out of 5 voters cast ballots for Obama. If 105,000 people voted, how many people voted for Obama?

_____

9. This year, a sports car is valued at $14,000, which is 7/8 of what its value was last year. Find the value of the sports car last year.

_____

10. Green's Plumbing charged $1480 for a water heater and installation. The charge for the heater was $1100, and Green's charged $20 per hour for installation. How many hours did the plumber work in order to finish the job?

_____

11. Kenji's monthly salary as a sales manager was $3660. This amount included his base monthly salary of $1000 plus commission of 4% on total sales. Find his total sales for the month.

          _____

12. The price of a pair of designer shoes at the Garcia Shoe Emporium is $500. This price includes the store's cost plus a mark-up rate of 20%. Find the Garcia Shoe Emporium's cost of the shoes.

          _____

13. The Roxy Dance Group rented the Paradise Theater for $3500. Additional expenses to put on the show amounted to $1255. How many tickets must be sold at $5.00 before the dance group makes a profit?

          _____

14. Philip is setting up a comic book store selling old comic books. The display cost $810, and each comic book costs $6.00. How many comic books would Philip have to sell at $15.00 each before he begins to make a profit?

          _____

15. An opera company rents a theatre for $6500. There is $3200 in additional expenses. How many tickets at $5.00 must be sold in order for the opera company to break even?

          _____

  a. 1300

  b. 1840

  c. 1940

  d. 640

  e. 1400

16. The sum of the squares of two consecutive positive integers is 145. Find the two integers.

_____

    a. 9, 10

    b. 8, 9

    c. 12, 13

    d. 10, 11

    e. 6, 7

17. Of the 60 psychologists at John Jay College, 30% are women. Without firing anyone, how many additional women must be hired in order to bring the total percentage up to 50% women?

_____

    a. 18

    b. 42

    c. 24

    d. 36

    e. 20

18. There are 80 mathematics professors at Borough of Manhattan Community College; 40% are men. Without firing anyone, how many additional men must be hired in order to bring the total percentage up to 50% men?

_____

    a. 32

    b. 16

    c. 48

    d. 24

    e. 8

19. Zelma and Eddie are reading the same book. Eddie reads 10 more pages per day than Zelma. Zelma can finish the book in 30 days; it takes Eddie 15 full days to finish the book. How many pages per day is Zelma reading?

_____

    a. 20

    b. 10

    c. 15

    d. 25

    e. 30

Chapter 18

Solving Inequalities

The same rules used to solve linear equations apply to solving inequalities with one exception: When dividing by a negative number, reverse the direction of the inequality sign.

Exercises

Write the correct answer in the space provided.

1. Solve the following inequality: $-3(x + 6) > -36$
_____

2. Solve the following inequality: $-3(x + 4) + 6x > 9$
_____

3. Solve the following inequality: $3x + 6 + 2x > 31$
_____

4. Solve: $4x - 7 > 2x + 9$
_____

5. Solve: $-7(x + 4) \leq 28$
_____

6. Solve: $-2x + 9 > -x + 4$ _____

7. Solve: $-x + 9 \leq 12$ _____

8. Saying that $3 < \sqrt{x} < 7$ is equivalent to saying what about x? _____

    a. $3 < x < 7$

    b. $8 < x < 14$

    c. $9 < x < 49$

    d. saying nothing about x

    e. $49 < x < 7$

9. Solve: $-3x + 6 \leq -9$ _____

10. Solve: $-x + 7 \geq 9$ _____

11. Solve: $3x + 6 + 2x > 66$ _____

12. Solve: $-2(x + 4) < 8$ _____

13. Solve: $6x - 7 \leq x + 8$ _____

Chapter 19

Factoring

Key Term

Factor   When two or more numbers are multiplied, each is called a factor. The process of finding all the factors used to produce an algebraic expression is called factoring.

Finding Common Factors

1. Find the largest number that divides evenly into the coefficients.

2. To factor out a variable as a common factor, the variable must appear in each part of the expression. Factor out the variables to the lowest exponent that appears.

3. Divide each term of the expression by the common factor.

Factoring Trinomials in Which the Coefficient of x^2 Is 1

For example, to factor

$$x^2 + bx + c$$

1. Factor out the highest common factor.

2. To factor, find two numbers whose product is the last term (c, the constant term) and whose sum is b, the middle term.

Factoring Trinomials in Which the Coefficient of x^2 Is Greater Than 1

1. Factor out the highest common factor.

2. List the positive factors of the coefficient as the first and last terms.

3. By trial and error find the correct combination of factors.

Factoring the Difference of Two Squares

1. The first term of the expression is a perfect square and the last term of the expression is a perfect square; the terms are separated by a negative sign.

2. Factor out the highest common factor.

3. Find the square root of the first term and then find the square root of the second term. One set of factors is separated by a positive sign and the second set of factors by a negative sign.

Exercises

Write the correct answers in the spaces provided. Factor completely.

1. Factor: $3x^3 - 6x^2 - 9x$ _____

2. Factor: $4x^2 - 36y^2$ _____

3. Factor: $x^2 + x - 12$ _____

4. Factor: $8x^6y^7 - 4x^2y^3 + 12xy$ _____

5. Factor: $x^2 - 7x + 10$ _____

6. Factor: $3x^3 - 21x^2 + 36x$ _____

7. Factor: $-7x^2y^3 + 14x^5y$ _____

8. Factor: $x^2 - 4y^2$ _____

9. Factor: $4x^2 - 64$ _____

10. Factor: $3x^3 - 9x^2 + 6x$ _____

11. Factor: $9x^2 - 16y^2z^2$ _____

12. Factor: $x^2 - 3x + 2$ _____

13. Factor: $x^2 + 8x - 9$ _____

14. Factor: $x^4 - y^4$ _____

15. Factor: $16A^4 - B^8$ _____

16. Which of the following is a factor of $x^2 - 5x - 6$? _____

    a. $x + 1$

    b. $x - 1$

    c. $x + 6$

    d. $x - 3$

    e. $x + 3$

17. Which of the following is a factor of $4x^3 + 6x^2 + 8x$? _____

    a. x

    b. $x + 2$

    c. $2x$

    d. x^2

    e. $x + 3$

Chapter 20

Rational Expressions

Key Terms

Fraction   Indicates the quotient of two numbers (the denominator cannot be zero).

Algebraic fraction   Indicates the quotient of two algebraic expressions. Also called "rational expression."

Reducing Algebraic Fractions

1. Completely factor numerator and denominator.

2. Divide the numerator and denominator by the greatest common factor (also known as "cancelling common factors").

Multiplication of Algebraic Fractions

1. Completely factor numerator and denominator.

2. Divide the numerator and denominator by the common factors.

Division of Algebraic Fractions

1. Use the multiplicative inverse (or reciprocal) of the divisor (the fraction after the division sign).

2. Change the division sign to multiplication; multiply the dividend by the reciprocal of the divisor.

3. Completely factor numerators and denominators.

4. Divide the numerators and denominators by the common factors.

Addition and Subtraction of Algebraic Fractions

1. Factor completely each denominator to determine the lowest common denominator (LCD).

2. Change each fraction to an equivalent fraction.

3. Add or subtract like terms in the numerator.

4. Reduce answer to lowest terms.

Exercises

Write the correct answers in the spaces provided.

1. For all $x \neq \pm 4$, $\dfrac{x^2 - 6x + 8}{x^2 - 16} = ?$ _____

2. For all $y \neq 3$ and $y \neq -4$, which expression is equivalent to $\dfrac{4y^2 + 16y}{y^2 + y - 12}$? _____

   a. $\dfrac{4y}{y - 3}$

   b. $\dfrac{4y}{y + 4}$

   c. $\dfrac{4y}{y - 4}$

   d. $\dfrac{y - 4}{y - 3}$

   e. $\dfrac{y - 4}{y + 4}$

3. Simplify: $\dfrac{x^2 - 5x + 4}{x - 1}$ _____

4. Simplify: $\dfrac{x - 5}{x^2 - 25}$ _____

5. Reduce $\dfrac{x^2 - 16}{x^2 + 6x + 8}$ to lowest terms. _____

6. Reduce $\dfrac{A^2 + 8A + 15}{A^2 + 5A + 6}$ to lowest terms.

_____

7. Multiply: $\dfrac{x - y}{3} \cdot \dfrac{9}{x - y}$

_____

8. Multiply: $\dfrac{2x + 1}{x^3} \cdot \dfrac{x^4}{8x + 4}$

_____

9. Multiply: $\dfrac{x^2 - 7x}{3x^2 - 27} \cdot \dfrac{x^2 - 9}{x^2 - 9x + 14}$

_____

10. Multiply: $\dfrac{x^2 - 1}{x^2 - 5x + 4} \cdot \dfrac{x + 3}{x + 1}$

_____

11. Divide: $\dfrac{a^2 + 14a + 49}{a + 7} \div \dfrac{a^2 - 49}{a - 7}$

_____

12. Divide: $\dfrac{x^2 + x - 2}{x^2 + 5x + 6} \div \dfrac{x - 1}{x}$

_____

13. Divide: $\dfrac{x^2 - 4}{x + 3} \div \dfrac{x^2 - 2x}{x}$

_____

14. Divide: $\dfrac{6 - 5x + x^2}{x^2 - 7x + 10} \div \dfrac{x^2 - 9x + 18}{x^2 - 8x + 15}$ _____

    a. $\dfrac{x - 3}{x - 2}$

    b. $\dfrac{x + 3}{x + 6}$

    c. $\dfrac{x - 3}{x - 6}$

    d. $\dfrac{x + 6}{x - 3}$

    e. $\dfrac{x - 5}{x - 2}$

15. Find the sum: $\dfrac{A + 4}{2A + 10} + \dfrac{5}{A^2 - 25}$ _____

16. Find the sum: $\dfrac{3}{x^2 + 6x + 8} + \dfrac{2}{x^2 - 16}$ _____

17. Find the sum: $\dfrac{y - 4}{y - 3} + \dfrac{2}{y^2 - 7y + 12}$ _____

18. Find the sum: $\dfrac{8x}{x^2 - 36} + \dfrac{2x}{x + 6}$ _____

19. Subtract: $\dfrac{3x + 2}{2x + 12} - \dfrac{x + 3}{x + 6}$ _____

20. Subtract: $\dfrac{6A + 5}{5A - 25} - \dfrac{A + 2}{A - 5}$ _____

21. Subtract: $\dfrac{3x}{x^2 - 81} - \dfrac{x - 4}{x + 9}$ _____

Chapter 21

Quadratic Equations

Any equation that can be put into the form $ax^2 + bx + c = 0$, $a \neq 0$, is called a quadratic equation. There are two solutions for each equation.

1. Set equation equal to zero.

2. Factor.

3. Set each factor equal to zero.

4. Solve for each of the factors.

Exercises

Write the correct answers in the spaces provided.

1. The solutions of $x^2 = 8x - 15$ are: _____

2. The solutions of $x^2 - 13x = -36$ are: _____

3. The solutions of $x^2 - 3x - 10 = 0$ are: _____

4. The solutions of $x^2 + 13x = -12$ are: _____

5. The solutions of $x^2 - 6x = -8$ are: _____

6. What is the sum of the roots of the quadratic equation $x^2 - 7x = -10$?   _____

    a. -7

    b. 7

    c. 11

    d. -11

    e. 10

7. What is the sum of the roots of the quadratic equation $x^2 - 4x = 5$?   _____

8. What is the sum of the roots of the quadratic equation $x^2 + 7x + 12 = 0$?   _____

9. What is the product of the roots of the quadratic equation $x^2 + 8x = -15$?   _____

10. What is the product of the roots of the quadratic equation $x^2 - 7x = 8$?   _____

11. What is the value of k when $\dfrac{x^2 - kx - 16}{x - 4} = x + 4$?   _____

    a. -1

    b. 1

    c. 0

    d. 4

    e. 5

12. What is the value of k when $\dfrac{x^2 - kx + 36}{x - 2k + 18} = x - 6$?   _____

Chapter 22

Simultaneous Equations

Simultaneous equations are two equations with two different variables wherein the variables' values are the same in both equations. You may use any of these methods to solve simultaneous equations.

Method #1

If the equations are in the following form,

$$x + y = 6$$
$$x - y = 2$$

1. Add the equations to eliminate one of the variables. Then solve for the other variable.

2. Substitute the value found for the variable in Step 1 in either equation and solve the equation for the other variable.

Method #2

If the equations are in the following form,

$$2x - y = 6$$
$$x + 3y = 3$$

1. Multiply each and every term on both sides of the first equation by 3 or multiply the second equation by –2 (depending on which variable you choose to eliminate).

2. Add the equations to eliminate one set of variables and solve for the other variable.

3. Substitute the value found for the variable in Step 2 in one of the equations and solve for the other variable.

| **Hint** | Methods #1 and #2 may both be called "Elimination." |

Method #3

Some simultaneous equations can be solved by substitution, for example

$$2x + 3y = 7$$
$$y = x - 1$$

1. In the above problem, substitute $x - 1$ for y in the top equation and then solve for x, as follows:

$$2x + 3(x - 1) = 7$$

2. Substitute the value found for the variable in either one of the equations and solve for the other variable.

Exercises

Write the correct answers in the spaces provided.

1. Solve: $A - B = 7$ and $2A + B = 2$ _____

    a. $A = -4$, $B = 3$

    b. $A = 3$, $B = 6$

    c. $A = 9$, $B = 2$

    d. $A = -3$, $B = 6$

    e. $A = 3$, $B = -4$

2. Solve the following system of equations: $3x + y = 8$ and $x = y$ _____

3. Solve the following system of equations: $2x - y = 6$ and $x + 3y = 3$ _____

4. Solve the following system of equations: $-2x + y = 6$ and $2x + 2y = 6$ _____

5. Solve: $2x - y = 8$ and $x + 3y = 4$ _____

6. Solve: $4x - y = 6$ and $x + 2y = 15$ _____

7. Solve: $2x + 3y = 7$ and $y = x - 1$ _____

8. Solve: $2x + 4y = 6$ and $x = y + 3$ _____

9. Solve: $2A - 3B = -3$ and $A + 4B = 4$ _____

10. Solve: $2x + y = 6$ and $y = x$ _____

11. The x-coordinate of the solution of the system $2x + 3y = -11$ and
$6x + y = 7$ is ? _____

Chapter 23

Graph of a Straight Line

Key Terms

Graph   A picture used to show information.

Origin   The center of the graph, where the x- and y-values are zero.

Slope   The ratio of the change in the vertical distance to the change in the horizontal distance.

x-axis   The horizontal line in a graph

x-intercept   The point where the line crosses the x-axis; where $y = 0$

y-axis   The vertical line in a graph

y-intercept   The point where the line crosses the y-axis; where $x = 0$

To draw a graph of a straight line, one must find points that are solutions to the equation for that line. For example, in the equation $x = y + 1$, one must assume values of one of the variables and solve for the other variable. If one assumes $y = 1$, substitute the value in the equation and solve for x; $x = 2$.

Graphing an Equation

Identify the points you plot with the following notation: (x, y). The x-value is written first, and then the y-value, separated by a comma; both are enclosed by parentheses. The point $(-3, -5)$ has an x-value of -3 and a y-value of -5.

1. Choose at least three different values for one variable in the equation $x = y + 1$.

2. Substitute in the equation and find the values of the other variable.

3. Plot the (x, y) values for each point on the graph at the right.

4. Connect the points as a straight line.

To Determine If a Point Lies on a Graphed Line

1. Substitute in the equation the values given.

2. If both sides of the equation have the same value, the point is a solution to the equation and lies upon the graphed line.

Slope-Intercept Form of the Equation of a Line

The slope-intercept form of the equation of the line with slope m and y-intercept b is:

$$y = mx + b.$$

1. Write the equation in the slope-intercept form:

$$y = mx + b.$$

2. Substitute in the equation the value of the slope and the value of the y-intercept.

To Find the Slope and the y-Intercept, Given an Equation

1. Solve the given equation for y and put it into the slope-intercept form of $y = mx + b$.

2. The slope is the number and sign in front of x, and the numeral and sign represented by the letter b is the y-intercept.

To Find the x-Intercept and the y-Intercept

To find the x-intercept (the point where the line crosses the x-axis, where $y = 0$), set $y = 0$ in the equation and solve for x.

To find the y-intercept (the point where the line crosses the y-axis, where $x = 0$), set $x = 0$ in the equation and solve for y.

To Write an Equation of a Straight Line
Given the x-Intercept and the y-Intercept

1. Use the formula

$$\frac{x}{a} + \frac{y}{b} = 1$$

where a is the value of the x-intercept and b is the value of the y-intercept.

2. Remove the fractions by using the lowest common multiplier (LCM).

3. Write the equation in the form $y = mx + b$.

To Find the Slope Given Two Points

Given any two points (x_1, y_1) and (x_2, y_2), the equation for the slope of the line on which they lie is

$$\text{Slope} = m = \frac{y_2 - y_1}{x_2 - x_1}$$

1. Label the points (x_1, y_1) and (x_2, y_2).

2. Write the formula for slope.

3. Substitute the values into the formula.

4. Evaluate to determine the slope m.

To Write an Equation Given the Slope and One Point

1. Use the point-slope formula

$$y - y_1 = m(x - x_1).$$

2. Label the point (x_1, y_1).

3. Substitute in the formula.

4. Use order of operations to solve for y.

5. Write the equation in the slope-intercept form of the equation of a straight line ($y = mx + b$).

Parallel and Perpendicular Lines

Parallel lines have equal slopes: $m_1 = m_2$.

Perpendicular lines: $m_1 = -\dfrac{1}{m_2}$

Hint	**To Find Perpendicular Lines**
	Use the inverse and the opposite sign of the original slope.
	$$y = 2x + 3 \text{ and}$$ $$y = -\frac{1}{2}x + 1$$
	are perpendicular lines.

Exercises

Write the correct answers in the spaces provided.

1. Find the slope of the line containing the points $(-3, -4)$ and $(-5, -6)$. _____

2. Find the *x*-intercept for the equation $2x - 3y = 6$. Write your answer as a point on the graph.

_____

3. Find the *y*-intercept for the equation $2x - 3y = 6$. Write your answer as a point on the graph.

_____

4. Find the slope of the line for the equation $3x + 5y = 15$.

_____

   a. 3

   b. $-\dfrac{3}{5}$

   c. -3

   d. $\dfrac{3}{5}$

   e. $\dfrac{5}{3}$

5. Find the slope of the line for the equation $3x - 7y = 12$.

_____

6. Which equation describes this table?

_____

x	1	2	-1
y	5	7	1

   a. $y = 2x + 3$

   b. $y = 3x + 1$

   c. $y = 2x - 3$

   d. $y = -2x + 5$

   e. $y = 3x + 5$

7. Which equation describes this table? _____

x	1	2	−1
y	4	7	−2

   a. $y = 3x - 7$

   b. $y = 3x + 1$

   c. $y = -3x + 1$

   d. $y = 2x - 2$

   e. $y = 2x + 2$

8. What is the equation of the line in the graph? _____

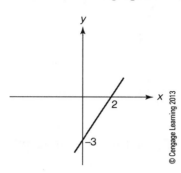

   a. $y = \dfrac{3}{2}x - 3$

   b. $y = \dfrac{3}{2}x$

   c. $y = \dfrac{-3}{2}x + 3$

   d. $y = \dfrac{3x - 3}{2}$

   e. $y = \dfrac{3x + 3}{2}$

9. Which equation has a slope perpendicular to the equation $2x + y = 6$? _____

   a. $y = 2x + \dfrac{1}{2}$

   b. $y = -2x + \dfrac{1}{2}$

   c. $y = 4x + 2$

   d. $y = -\dfrac{1}{2}x + 2$

   e. $y = \dfrac{1}{2}x + 2$

10. Which equation has a slope perpendicular to the equation $-3x + y = 7$?   _____

    a. $y = -\dfrac{1}{3}x + 7$

    b. $y = 3x + 4$

    c. $y = \dfrac{1}{3}x + 6$

    d. $y = 2x - 4$

    e. $y = 2x + 4$

11. If the slope of a line is 6 and the y-intercept is -12, what is the x-intercept?   _____

    a. 2

    b. -2

    c. -12

    d. 12

    e. 4

12. If the slope of a line is 8 and the y-intercept is 12, what is the x-intercept?   _____

13. Write an equation of a straight line whose slope is 2 and that passes through the point $(-3, 4)$.   _____

    a. $y = 2x + 10$

    b. $y = 2x - 10$

    c. $y = 2x + 2$

    d. $y = 2x - 2$

    e. $y + 2x = -10$

14. Write an equation of a straight line that is parallel to $y + 3x = 9$ and passes through the point $(2, 7)$.   _____

    a. $y = 3x - 13$

    b. $y = -3x + 2$

    c. $y = -3x - 13$

    d. $y = -3x + 13$

    e. $y = 2x - 10$

15. Write an equation in slope-intercept form of a straight line whose slope is -3 and that passes through the point $(-4, -6)$. _____

16. Write an equation in slope-intercept form of a straight line that is perpendicular to $y + 4x = 9$ and passes through the point $(-4, 8)$. _____

17. Write the slope-intercept form of an equation of a straight line whose slope is -2 and whose y-intercept is -3. _____

18. What is the slope of a straight line whose x-intercept is -3 and whose y-intercept is 5? _____

19. Write an equation of a straight line whose x-intercept is 2 and whose y-intercept is 3. _____

20. Which of the following points lies on the graph? _____

$$2x + y = 6$$

a. $(2, 2)$

b. $(-2, -2)$

c. $(3, 6)$

d. $(0, 2)$

e. $(-1, 1)$

21. Which one of the following points lies on the graph?

$$2x - 3y = 6$$

    a. (3, 4)

    b. (2, 1)

    c. (–1, –1)

    d. (3, 0)

    e. (2, –1)

22. What is the slope of a line that contains the points with (x, y) coordinates (–7, 2), (4, –6)?

Chapter 24

Radicals

To Simplify Radicals

1. Find two factors of the numeral, one of which is the largest perfect square.

2. To take the square root of a variable with an even exponent, divide the exponent by 2. If the variable has an odd exponent, make it even by reducing the exponent by 1. For example, $x^{13} = x^{12} \times x$ or $y^{15} = y^{14} \times y$.

Addition and Subtraction of Radicals

1. Simplify each of the radicals.

2. Add or subtract the coefficients in front of like radicals.

Multiplication and Division of Radicals

1. Multiply or divide the coefficients.

2. Multiply or divide the radicands.

3. Simplify answer to lowest terms.

To Rationalize an Irrational Radical Denominator

1. Multiply the numerator and the denominator by the radical in the denominator (in essence, multiply the fraction by 1). The resulting denominator becomes a rational number. If the denominator is a binomial, multiply the numerator and denominator by the binomial with the sign in the middle changed to the opposite.

2. Reduce answer to lowest terms.

Exercises

Write the correct answers in the spaces provided.

1. Simplify: $\sqrt[3]{125} + \sqrt[3]{27}$ _____

2. Simplify: $\sqrt{8x^{23}}$ _____

3. Simplify the radical expression: _____

   $\sqrt[4]{81} + \sqrt[4]{16}$

   a. 5

   b. 7

   c. 9

   d. 6

   e. 8

4. Simplify: $12\sqrt{8} + 4\sqrt{32} - 8\sqrt{50}$ _____

5. Simplify: $\sqrt{36x^{19}y^{21}}$ _____

6. Simplify: $\sqrt{6x^2y^2} \cdot \sqrt{8x^3y^6}$ _____

7. Simplify: $\sqrt{9x^2y^4} \cdot \sqrt{8xy^2}$ _____

   a. $6xy^3\sqrt{2x}$

   b. $2x\sqrt{6xy^3}$

   c. $2y\sqrt{3x^3}$

   d. $2x\sqrt{6x^2y^3}$

   e. $6x^2y^3\sqrt{2x}$

8. Simplify: $\sqrt{72x} + 6\sqrt{98x} - 12\sqrt{8x}$ _____

   a. $-24\sqrt{2x}$

   b. $24\sqrt{2x}$

   c. $12\sqrt{2x}$

   d. $7\sqrt{2x}$

   e. $-7\sqrt{2x}$

9. Simplify: $\sqrt{9xy^7} \cdot \sqrt{8x^2y^9}$ _____

   a. $4xy^8\sqrt{2x}$

   b. $\sqrt{72x^3y^{16}}$

   c. $6x\sqrt{2xy^4}$

   d. $6xy^4\sqrt{2x}$

   e. $6xy^8\sqrt{2x}$

10. Solve: $\sqrt{x+4} - 2 = 4$ _____

   a. 64

   b. 32

   c. 8

   d. 16

   e. 4

11. Simplify: $\sqrt{x^2 + 8x + 16}$ _____

   a. $x - 4$

   b. $x + 8$

   c. $x + 4$

   d. $x - 8$

   e. $x + 6$

12. Simplify: $\sqrt{x^2 + 6x + 9}$ _____

   a. $x + 3$

   b. $x - 3$

   c. $x + 9$

   d. $x + 1$

   e. $x - 9$

13. Rationalize: $\dfrac{3 + \sqrt{7}}{3 - \sqrt{7}}$ _____

14. Rationalize: $\dfrac{2 + \sqrt{5}}{2 - \sqrt{5}}$ _____

15. Simplify: $\dfrac{10}{\sqrt{11} - 8}$ _____

   a. $\dfrac{-80 - 10\sqrt{11}}{53}$

   b. $\dfrac{80 + 11\sqrt{11}}{100}$

   c. $\dfrac{80 - 10\sqrt{11}}{100}$

   d. $\dfrac{80 + 10\sqrt{11}}{57}$

   e. $\dfrac{80 + 11\sqrt{10}}{53}$

16. Solve: $\sqrt{a^2 + 88} = a + 22$ _____

   a. -19

   b. 1

   c. -14

   d. -9

   e. -12

17. Find the solution to the following equation: _____

$$3 + \sqrt{5 - x} = 6$$

    a. 3

    b. 4

    c. −3

    d. −4

    e. 0

18. Simplify: $3x\sqrt{xy^2} + \sqrt{64x^2y^3}$ _____

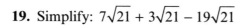

19. Simplify: $7\sqrt{21} + 3\sqrt{21} - 19\sqrt{21}$ _____

20. Simplify: $\sqrt{x^2 + 12x + 36}$ _____

21. Solve: $\sqrt{7w} - 12 = 19$ _____

22. Simplify: $\sqrt{81x^{16}y^{18}}$ _____

23. Simplify: $8x\sqrt{75y} - 2x\sqrt{20y} + 2x\sqrt{45y}$ _____

24. Simplify: $\dfrac{\sqrt{3x}}{\sqrt{x}}$ _____

25. Solve: $\sqrt{x+4} = 2 - x$ _____

An Introduction to Geometry

Parallelogram   A quadrilateral in which both pairs of opposite sides are parallel.

Polygon   A closed figure formed by three or more straight lines, all of which intersect in a plane.

Quadrilateral   A polygon with four sides.

Rectangle   A parallelogram in which all angles are right angles.

Rhombus   A parallelogram in which all sides are equal in length.

Square   A rectangle in which all sides are equal.

Trapezoid   A quadrilateral in which only one pair of opposite sides are parallel.

Triangle   A polygon with three sides and three angles.

Geometry is the study of the shapes, sizes, and relationships of figures such as squares, rectangles, triangles, and circles.

Important geometric concepts include the following.

Angle Theorems

Supplementary angles   Supplementary angles are two or more angles that add up to 180°.

Complementary angles   Complementary angles are two or more angles that add up to 90°.

Vertical angles   Vertical angles are formed when two lines intersect. The opposite angles are equal. They are called vertical angles.

Triangles: Properties and Formulas

A triangle is a polygon with three sides and three angles. (A polygon is a closed figure formed by three or more sides.) The symbol for a triangle is △. The sum of the angles of a triangle is 180°.

Equilateral triangle   A triangle with three sides of equal length.

Isosceles triangle   A triangle with two sides of equal length.

Scalene triangle   A triangle with no two sides of equal length.

Right triangle   A triangle with a right (90°) angle.

Pythagorean Theorem

This theorem applies only to a right triangle, which contains a 90° angle. The side opposite the right angle is called the hypotenuse. The other two sides are called legs. The hypotenuse is the longest side.

This theorem states:

$$\text{Leg}^2 + \text{Leg}^2 = \text{Hypotenuse}^2$$
$$a^2 + b^2 = c^2$$

Polygons: Properties and Formulas

Area   The amount of the plane enclosed by a polygon.

Perimeter   The distance around the polygon.

Area of a Rectangle

Area = length × width

$$A = l \times w$$

Area of a Square

Area = side squared

$$A = s^2$$

Area of a Triangle

Area = $\frac{1}{2}$ base × height

$$A = \frac{1}{2}b \times h$$

Area of a Trapezoid

Area = $\frac{1}{2}$ height × (sum of the bases)

$$A = \frac{1}{2}h(b_1 + b_2)$$

Perimeter of a Rectangle

Perimeter = 2 × length + 2 × width

$$P = 2l + 2w$$

Perimeter of a Square

Perimeter = 4 times the side

$$P = 4s$$

Circles: Properties and Formulas

Circle   A plane figure bounded by a curved line, every point of which is the same distance from the center of the figure. Following are key parts of a circle:

Circumference   The line that forms its outer boundary. It is like the perimeter of a circle.

Diameter   A line segment joining two points on the circumference and passing through the center. A diameter is equal to two radii.

Pi (π)   The ratio of a circle's circumference to its diameter.

Radius   A line segment joining the center to any point on the circumference.

Formulas for Circles

Circumference = pi × the diameter or pi × (the radius × 2)

$$C = \pi d$$

or

$$C = 2\pi r$$

Area = pi × the radius squared

$$A = \pi r^2$$

Diameter = the radius × 2

$$d = 2r$$

Pi = the ratio between the circumference and the diameter of a circle

$$\pi \approx \frac{22}{7} \approx 3.14$$

Exercises

Write the correct answers in the spaces provided.

1. Find the value of x.   _____

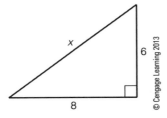

2. Find the length of a rectangle whose width is $3xy$ yards and whose area is
   $6x^2y + 21x^2y^2$ square yards.   _____

   a. $2x^3 + 7xy$

   b. $2x + 7xy$

   c. $2x + 7x^2y^2$

   d. $2x^2 + 7xy$

   e. $2x + 7xy^2$

3. Find the area of a rectangle whose length is $3x^2y^2$ yards and whose width is
   $2xy$ yards.   _____

4. The area of a square is $16x^4y^5$ square feet. Find the measure of its side.   _____

5. Find the perimeter of a rectangle whose length is $4x^2y^2$ feet and whose
   width is $3xy$ feet.   _____

6. Find the area of a rectangle whose length is 6 feet more than twice its width.   _____

7. Find the width of a rectangle whose perimeter is $36x^2y^2$ feet and whose length is $12x^2y^2$ feet.

_____

8. Find the perimeter of a rectangle whose width is 2 feet less than the length.

_____

9. Find the perimeter of a square whose area is $64A^2$ square yards.

_____

10. Find the area of a square whose perimeter is $100xy$ feet.

_____

11. Find the value of the missing side.

_____

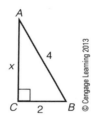

12. Find the value of the missing side.

_____

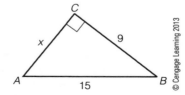

13. The radius of a circle is $2x$ centimeters. Find the diameter. _____

14. The radius of a circle is $3xy$ centimeters. Find the circumference (answer can be left in terms of π). _____

15. Find the area of a circle whose diameter is $6x$ centimeters (answer can be left in terms of π). _____

16. The circumference of a circle is 12π centimeters. Find the diameter. _____

Practice Placement Exams

Exam A

1. If $x = -2$ and $y = 3$, what is the value of the product $3(x + y)(x - y)$? _____

   a. 15

   b. -15

   c. 5

   d. -5

   e. 75

2. What is the product of $(w + 2x)$ and $(y - z)$? _____

   a. $2xy + wz - 2xz + 2xy$

   b. $wz - 2xz + 2xy + wy$

   c. $wz + 2xz - 2xy - wy$

   d. $wy - 2xy - wz + 2xz$

   e. $wy + 2xy - wz - 2xz$

3. Each week, Marcia gets paid D dollars plus a 9% commission on her total sales S. What is Marcia's weekly pay? _____

   a. $D + 0.09S$

   b. $8D + S$

   c. $S + 0.09D$

   d. $0.09(D + S)$

   e. $D + 8S$

4. Solve the following equation: $2(3 + A) - 3(5 + A) = 13$ _____

    a. 15

    b. 5

    c. –4

    d. –12

    e. –22

5. For all $x \neq 0$ and $y \neq 0$, $\dfrac{x^{-3}y^2}{x^5 y^{-4}} = ?$ _____

    a. $\dfrac{x^2}{y^2}$

    b. $\dfrac{y^3}{x^4}$

    c. $\dfrac{y^6}{x^2}$

    d. $\dfrac{y^6}{x^8}$

    e. $\dfrac{1}{x^2 y}$

6. For all x, y, and z, $(-3x^3 y^2 z)^2 = ?$ _____

    a. $-6x^5 y^4 z^2$

    b. $-9x^6 y^4 z^2$

    c. $9x^6 y^4 z^2$

    d. $6x^5 y^4 z^2$

    e. $9xyz$

7. Solve: $3(2y + 5) - 4(y - 2) = 3(2 + 2y) + 1$ _____

    a. $y = -5$

    b. $y = 4$

    c. $y = 9$

    d. $y = 0$

    e. $y = 3$

8. Find the product of $(x^3 + 2x^2 - 2x + 3)$ and $(x - 5)$.

a. $x^4 - 3x^3 - 12x^2 + 13x - 15$

b. $x^4 + 7x^3 + 12x^2 + 13x + 15$

c. $x^4 - 3x^3 - 12x^2 - 7x - 15$

d. $x^4 + 2x^3 - 2x^2 + 3x$

e. $x^4 + x^3 - 2x^2 + 3x$

9. Write an equation for the line.

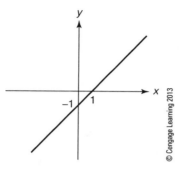

a. $x - 1 = 1$

b. $x - y = 1$

c. $x = y + 2$

d. $x + y = 1$

e. $x = y - 2$

10. There are 80 mathematics professors at John Jay College; 25% are women. Without firing anyone, how many additional women must be hired in order to bring the total percentage up to 50% women?

a. 40

b. 32

c. 20

d. 60

e. 50

11. This is a graph of which equation?

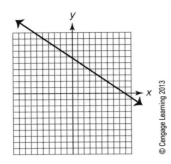

    a. $y = -\dfrac{3}{2}x + 6$

    b. $y = \dfrac{3}{2}x + 6$

    c. $y = \dfrac{2}{3}x + 6$

    d. $y = -\dfrac{2}{3}x + 6$

    e. $y = -\dfrac{2}{3}x - 6$

12. Simplify: $\dfrac{6 - 5x + x^2}{x^2 - 7x + 10} \div \dfrac{x^2 - 9x + 18}{x^2 - 8x + 15}$

    a. $\dfrac{x - 3}{x - 2}$

    b. $\dfrac{x - 3}{x - 6}$

    c. $\dfrac{x + 3}{x + 6}$

    d. $\dfrac{x - 5}{x - 2}$

    e. $\dfrac{x + 6}{x - 3}$

Exam B

1. Which equation represents the line that contains the points (4, 6) and (2, 3)? _____

   a. $y = \dfrac{3}{2}x$

   b. $y = \dfrac{3}{2}x + 6$

   c. $y = -\dfrac{3}{2}x$

   d. $y = \dfrac{2}{3}x + 6$

   e. $y = \dfrac{2}{3}x$

2. For all $x \neq \pm 5$, $\dfrac{x^2 + 7x + 10}{x^2 - 25} = ?$ _____

   a. $\dfrac{x + 2}{x + 5}$

   b. $\dfrac{x - 2}{x - 5}$

   c. $\dfrac{(x + 6)(x + 1)}{(x + 5)(x - 5)}$

   d. $\dfrac{x - 2}{x + 5}$

   e. $\dfrac{x + 2}{x - 5}$

3. On three of four mathematics tests, Guillermo received grades of 87, 91, and 81. What must Guillermo's grade be on the fourth test to have a mean (average) of exactly 85? _____

   a. 85

   b. 87

   c. 82

   d. 81

   e. 83

4. Which of the following expressions represents the product of 5 more than twice a number and 7 less than the number?

    a. $2x^2 - 19x - 35$

    b. $2x^2 + 9x - 35$

    c. $2x^2 - 9x - 35$

    d. $2x^2 + 19x - 35$

    e. $2x^2 - 19x + 35$

5. If $c = -2$ and $d = -3$, what is the value of $4c^3 - 2cd$?

    a. -44

    b. -36

    c. -20

    d. 20

    e. 36

6. What is the value of $27^{\frac{2}{3}}$?

    a. 18

    b. 3

    c. 6

    d. 9

    e. 12

7. For all x, $(x^2 + 3x + 9)(x - 3)$ is which of the following?

    a. $x^3 - x^2 + 3x - 3$

    b. $x^3 - 27$

    c. $3x - 3$

    d. $x^2 + 2x - 3$

    e. $x^3 + 3x^2 - x - 3$

8. What is the sum of the roots of the quadratic equation $x^2 + 6x = -8$?

    a. 6

    b. 9

    c. 8

    d. -6

    e. -9

9. What is the slope of the line $2x - y = 7$?

    a. -7

    b. 7

    c. 2

    d. -2

    e. $-\dfrac{1}{2}$

10. Simplify: $\dfrac{\sqrt{75}}{3} + \dfrac{\sqrt{12}}{6}$

    a. $2\sqrt{3}$

    b. $\dfrac{\sqrt{87}}{9}$

    c. $\dfrac{12\sqrt{3}}{3}$

    d. $\dfrac{10\sqrt{3}}{6}$

    e. $\dfrac{10\sqrt{3}}{9}$

11. Solve: $3^{3n+1} = 9^{n-7}$

    a. 0

    b. -6

    c. 12

    d. 15

    e. -15

12. Which equation describes the data in this table?

x	-2	0	2
y	9	5	1

    a. $y = x - 6$

    b. $y = 4x - 7$

    c. $y = 4x + 3$

    d. $y = 3x + 5$

    e. $y = -2x + 5$

Exam C

1. Which expression represents the product of 3 more than a number x and 2 less than twice x?

   a. $2x^2 + 8x + 6$

   b. $2x^2 + 4x - 6$

   c. $2x^2 - 4x + 6$

   d. $2x^2 + 8x - 6$

   e. $2x^2 - 4x - 6$

2. For all $x \neq \pm 3$, $\dfrac{x^2 + 9x + 18}{x^2 - 9} = ?$ _____

   a. $\dfrac{x - 6}{x - 3}$

   b. $\dfrac{x + 6}{x + 3}$

   c. $\dfrac{x + 9}{x + 6}$

   d. $\dfrac{x + 6}{x - 3}$

   e. $\dfrac{x - 6}{x + 3}$

3. What are the solutions of the quadratic equation $x^2 - 7x = -12$? _____

   a. 3 and –4

   b. –2 and –6

   c. 2 and 6

   d. 3 and 4

   e. –3 and –4

4. What is the sum of the solutions of the quadratic equation $y^2 + 3y = 28$?

   a. 10

   b. 11

   c. –11

   d. –3

   e. 3

5. If $c = -1$ and $d = -2$, what is the value of the expression $2c^2d - 3cd$? _____

   a. -10

   b. -24

   c. 2

   d. 10

   e. -2

6. Takao scores a 90, an 84, and an 89 on three out of four math tests. What must Takao score on the fourth test to have an 87 average (mean)? _____

   a. 87

   b. 88

   c. 85

   d. 84

   e. 86

7. Simplify: $8\sqrt{17} - 2\sqrt{17} + 7\sqrt{17}$ _____

   a. $17\sqrt{13}$

   b. $9\sqrt{17}$

   c. $6\sqrt{17}$

   d. $13\sqrt{17}$

   e. $42\sqrt{17}$

8. Simplify: $\sqrt{x^2 + 18x + 81}$ _____

   a. $x - 9$

   b. $\sqrt{x} + \sqrt{9}$

   c. $x + 81$

   d. $x + 9$

   e. $\sqrt{x + 9}$

9. Simplify: $(-2x^3)(x^{-5})^2$ _____

    a. $\dfrac{2}{x^7}$

    b. $-\dfrac{1}{2x^7}$

    c. $2x^7$

    d. $-\dfrac{2}{x^7}$

    e. $-\dfrac{2}{x^5}$

10. An exit poll showed that 7 out of every 9 voters cast a ballot in favor of Hillary Clinton for president. At this rate, how many voters voted in favor of Hillary Clinton if 18,000 people voted? _____

    a. 2000

    b. 12,000

    c. 9000

    d. 13,000

    e. 14,000

11. Solve: $\dfrac{4x}{3} = \dfrac{x-14}{6}$ _____

    a. 2

    b. −9

    c. −2

    d. −1

    e. 7

12. Write the equation of a straight line that is parallel to $y + 3x = 6$ and that passes through the point $(3, -4)$. _____

    a. $y = 3x - 13$

    b. $y = -3x + 13$

    c. $y = 4x + 6$

    d. $y = -3x - 5$

    e. $y = -3x + 5$

Exam D

1. Simplify: $(a + b)^2 - (7ab - 8a^2)$ _____

   a. $9a^2 - 5ab + b^2$

   b. $9a^2 + 5ab + b^2$

   c. $8a^2 - 5ab + b^2$

   d. $8a^2 + 5ab + b^2$

   e. $9a^2 - 9ab + b^2$

2. Solve the following linear inequality: $-3y + 6 < 12$ _____

   a. $y > -6$

   b. $y < -6$

   c. $y < 2$

   d. $y > -2$

   e. $y > -9$

3. For all $x \neq \pm 2$, $\dfrac{x^2 - 5x + 6}{x^2 - 4} = ?$ _____

   a. $\dfrac{x - 2}{x + 2}$

   b. $\dfrac{x + 3}{x - 2}$

   c. $\dfrac{x + 3}{x + 2}$

   d. $\dfrac{x - 3}{x + 2}$

   e. $\dfrac{x - 2}{x + 3}$

4. Evaluate the expression for A: _____

   $$A = P(1 + r) \text{ when } P = \$550 \text{ and } r = 8\%$$

   a. $594

   b. $494

   c. $549

   d. $484

   e. $44

5. Simplify the radical expression: $\sqrt[3]{64} + \sqrt[3]{125}$ _____

    a. 1

    b. 17

    c. 14

    d. 13

    e. 9

6. Simplify the following: $\dfrac{24x^6 - 6x^4 + 4x^2}{4x^2}$ _____

    a. $6x^8 - 2x^6$

    b. $6x^8 - 2x^6 + 1$

    c. $6x^4 - \dfrac{3}{2}x^2$

    d. $6x^4 - \dfrac{3}{2}x^2 + 1$

    e. $8x^6 + 1$

7. What is the equation of the line that contains the points with (x, y) coordinates $(-3, 7)$ and $(5, -1)$? _____

    a. $y = x + 10$

    b. $y = 3x - 2$

    c. $y = -\dfrac{3}{4}x + \dfrac{11}{4}$

    d. $y = -x + 4$

    e. $y = -\dfrac{1}{3}x + 8$

8. Write an equation of the line whose slope is -2 and that passes through the point $(-4, 5)$. _____

    a. $y = 2x + 3$

    b. $y = -3x + 2$

    c. $y = 3x - 2$

    d. $y = -2x + 3$

    e. $y = -2x - 3$

9. If $a + bx = c$, then $x = ?$ _____

    a. $\dfrac{ac}{b}$

    b. $\dfrac{c}{b} - a$

    c. $\dfrac{c - a}{b}$

    d. $\dfrac{a - c}{c}$

    e. $\dfrac{ab}{c}$

10. Which of the following is a factor of $x^2 - 7x - 18$? _____

    a. $x - 2$

    b. $x + 9$

    c. $x - 9$

    d. $x + 3$

    e. $x + 6$

11. Perform the indicated operation. $\sqrt{8xy^2} \cdot \sqrt{6x^2y^4}$ _____

    a. $4xy^3\sqrt{3x}$

    b. $2xy^3\sqrt{3x}$

    c. $4xy^3\sqrt{12x}$

    d. $2y\sqrt{3x^3}$

    e. $2x^2y^3\sqrt{3x}$

12. The World Music Store set up a display of CDs. The display cost $540, and each CD cost $5.00. How many CDs must the World Music Store sell at $15.00 each before it makes a profit? _____

    a. 36

    b. 54

    c. 90

    d. 18

    e. 72

Exam E

1. For all x, $(x - 4)^2 + 3(2x - 4) = ?$

    a. $2x^2 - 2x + 4$

    b. $2x^2 + 2x - 16$

    c. $2x^2 + 2x - 4$

    d. $x^2 + 2x + 4$

    e. $x^2 - 2x + 4$

2. What is the solution of the system of equations $x + 2y = 7c$ and $3x - 2y = 5c$?

    a. $(c, 2c)$

    b. $(5c - c)$

    c. $(3c, -2c)$

    d. $(-3c, 2c)$

    e. $(3c, 2c)$

3. What is the value of k when $\dfrac{x^2 - kx + 24}{x - 12} = x - 2$?

    a. 2

    b. -14

    c. -10

    d. 14

    e. 10

4. Solve: $64^{-\frac{2}{3}} = ?$

    a. 16

    b. $\dfrac{1}{16}$

    c. -16

    d. $-\dfrac{1}{16}$

    e. $\dfrac{1}{4}$

5. Find the length of a rectangle whose width is $4xy$ and whose area is $16xy^2 + 24x^2y$.

    a. $10xy$

    b. $4y + 6x$

    c. $12y + 20x$

    d. $24xy$

    e. $10y + 10x$

6. If the slope of a line is 6 and the y-intercept is -12, what is the x-intercept?

    a. -2

    b. -4

    c. $\dfrac{1}{2}$

    d. 4

    e. 2

7. Solve the following set of equations: $3x - 7y = -4$ and $2x - 5y = -3$

    a. $(1, 2)$

    b. $(2, 1)$

    c. $(-2, -1)$

    d. $(1, 1)$

    e. $(-1, -1)$

8. Find the solution of the following equation: $7 - \sqrt{4 - x} = 4$

    a. -13

    b. 13

    c. -5

    d. 5

    e. 6

9. Which one of the following points lies on the graph of $7x - 6y = 1$?

    a. $(-1, 4)$

    b. $(2, 9)$

    c. $(-2, -7)$

    d. $(3, 7)$

    e. $(1, 1)$

10. Write the length of the missing side in the space provided.   _____

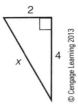

© Cengage Learning 2013

   a. $2\sqrt{5}$

   b. 20

   c. $5\sqrt{2}$

   d. 6

   e. $\sqrt{20}$

11. A map is drawn so that 2 inches represents 700 miles. If the distance between two cities is 3850 miles, how far apart are they on the map?   _____

   a. 5.5 inches

   b. 11 inches

   c. 22 inches

   d. 6 inches

   e. 12 inches

12. Simplify: $\sqrt{396x} + 3\sqrt{11x} - 12\sqrt{275x}$ _____

   a. $-49\sqrt{11x}$

   b. $49\sqrt{11x}$

   c. $-51\sqrt{x}$

   d. $-49\sqrt{x}$

   e. $-51\sqrt{11x}$

Answer Key

Part I   Arithmetic

Chapter 1
Numerals

1. c
2. e
3. a
4. six million four hundred thousand five
5. three hundred twenty-five thousand one hundred nine
6. six billion one hundred ninety-five thousand seventy-one
7. d
8. a
9. e
10. e
11. 9,000,020,003
12. 3,000,106

Chapter 2
Adding and Subtracting Units

1. 1 hour, 20 minutes
2. 3 hours, 24 minutes
3. 2 yards 4 inches
4. 7 yards 1 foot 1 inch
5. 3 hours, 37 minutes
6. 5 yards 1 foot
7. $5.70
8. $12.85
9. $1.69
10. 1 hour, 30 minutes
11. 62,040
12. 28
13. 49 feet
14. b
15. a
16. e
17. a
18. 3 yards 2 feet

Chapter 3
Fractions

1. a
2. b
3. $9\frac{5}{14}$
4. $\frac{2}{5}$
5. $4\frac{1}{12}$
6. $-\frac{1}{3}$
7. $-2\frac{2}{9}$
8. $13\frac{4}{9}$
9. 2
10. b, c
11. a, d
12. b
13. e
14. $\frac{4}{9}$
15. $\frac{1}{5}$
16. 48
17. c
18. e
19. $\frac{5}{12}$
20. 5000
21. $\frac{1}{20}$
22. $\frac{9}{16}$

Chapter 4
Decimals

1. d
2. hundredths
3. ten-thousandths
4. a
5. a
6. b
7. d
8. e
9. a
10. d
11. 2.321
12. 5.025
13. 0.01
14. 100
15. 0.0001
16. 0.0038
17. d
18. e
19. 30 toys
20. 0.55 pound
21. $8,445
22. 46 toys
23. d
24. d
25. d
26. 4.01, 3.333, 3.1, 3.099
27. 3.099, 3.1, 3.333, 4.01
28. b
29. a
30. b
31. e
32. $51.24
33. $26.60
34. 17.75 ft
35. 35.4 ft
36. 12
37. c
38. d
39. d

Chapter 5
Conversions: Decimals, Fractions, and Percents

1. 36%
2. 35%
3. $\dfrac{11}{200}$
4. 212.5%
5. 0.037
6. 70%
7. $\dfrac{1}{125}$
8. $1\dfrac{7}{20}$

9. b
10. b
11. $7\dfrac{17}{20}$
12. 3.45%
13. 3.64%
14. $377\dfrac{7}{9}\%$
15. 0.16
16. 20%
17. 0.35
18. $\dfrac{7}{20}$
19. $\dfrac{1}{50}$
20. $\dfrac{37}{100}$
21. $1\dfrac{9}{20}$
22. 0.39
23. 1.45
24. 3.1

Chapter 6
Rounding Off

1. 0.43
2. c
3. d
4. b
5. c
6. 14.3%
7. 12.71
8. $2.53
9. 0.58
10. 0.6
11. $66.33
12. 91.7
13. 9.6
14. 16.27

Chapter 7
Percent Problems

1. 500
2. 78
3. 14.5
4. e
5. b
6. 18
7. 80% of 16
8. a
9. d
10. 99%
11. a

12. d
13. e
14. d
15. b
16. 19.25%
17. 6% of 180
18. 30
19. 360
20. $11.36
21. 225
22. 3136
23. 8.64
24. 600
25. $2.85
26. $351.81
27. $25.50
28. c

Chapter 8
Scientific Notation

1. e
2. b
3. c
4. c
5. c
6. d
7. b
8. 4270
9. 0.0934
10. 12,700
11. 0.0064
12. 7.38×10^{6}
13. 7.59×10^{-3}
14. 3.739×10^{6}
15. 3.51×10^{5}

Chapter 9
Statistics

1. a
2. a
3. b
4. b
5. a
6. b
7. 3
8. 7
9. 3.67
10. 45
11. $301
12. $10.75
13. 3, 2
14. 1.5

15. 48.9 miles per hour
16. 91
17. d

Chapter 10
Area, Perimeter, and Cost

1. 40 square yards
2. 1950 square inches
3. $504
4. $315
5. 30 ft
6. 110 ft
7. $840
8. $378
9. $864
10. 12 in.
11. c
12. a
13. b

Chapter 11
Ratios and Proportions

1. d
2. e
3. a
4. a
5. b
6. c
7. b
8. b
9. e
10. b
11. a
12. b
13. c
14. a
15. a
16. d
17. c
18. d
19. b
20. d
21. a
22. c
23. e
24. a
25. e
26. $18.30 per minute
27. a
28. 180 miles
29. b
30. $x = -\dfrac{8}{3}$

Part II Algebra and Geometry

Chapter 12
Rules for Signed Numbers

1. 36°C
2. 15
3. $700
4. $197\frac{1}{4}$ pounds
5. −7
6. 30
7. 31
8. 7
9. 144
10. −243
11. 78
12. −169
13. −48
14. 26
15. 847
16. 131
17. 1
18. 249
19. 1721
20. e
21. b
22. $180 - N$
23. $75h$
24. $\dfrac{m}{x}$
25. $10d + 25q$
26. b
27. $x(x + 5)$ or $x^2 + 5x$
28. $x^2 + 6x$ or $x(x + 6)$
29. $\dfrac{x}{x - 12}$
30. $5n + 25q$

Chapter 13
Exponent Rules

1. x^7
2. $x^5 y^5$
3. $\dfrac{1}{x^6}$
4. $\dfrac{1}{x^2}$
5. x^2
6. $\dfrac{y^8}{x^6}$
7. $\dfrac{y^3}{x^2}$

8. $\dfrac{1}{y^4}$
9. 1
10. $24x$
11. $-\dfrac{18}{x^4}$
12. e
13. d
14. c
15. a
16. $48x$
17. $16y$
18. $\dfrac{7}{17x^{16}B^{17}}$
19. d
20. a
21. b
22. b
23. c
24. c
25. e
26. b
27. b
28. e
29. a
30. a
31. e
32. c
33. b
34. e
35. e
36. e
37. b
38. e
39. d

Chapter 14
Operations with Polynomials

1. $-3x^3y - 10x^2y^3$
2. $9x^3 - 7x^2 + 3x - 6$
3. $-2y^2 + 3xy + x$
4. $2x^2 - 14$
5. $4x + 14$
6. $-x - 9$
7. $14r^2s^2t^2$
8. $-27x^6y^{12}$
9. $-192x^{17}y^{13}$
10. $\dfrac{243x^7}{y^{10}}$
11. $-x^2$
12. e

13. $2x^2 - 13x + 21$

14. $x^2 - 14x + 49$

15. c

16. e

17. $x^3 - 3x^2 - 9x + 27$

18. e

19. $3x^5 + x^3 + 1$

20. $8x^2 - 4x - 24$

21. b

22. a

23. $\dfrac{y-2}{2y+3}$

24. $\dfrac{x-3}{x+4}$

Chapter 15
Evaluating Algebraic Expressions and Formulas

1. -20

2. 6

3. c

4. 1

5. e

6. -3

7. $\dfrac{7}{3}$

8. 12

9. 8

10. b

11. d

12. 40 feet

13. $32

14. 12 feet

15. 36 square feet

Chapter 16
Linear Equations

1. $x = 5$

2. $x = -46$

3. $x = 22$

4. $x = 0$

5. a

6. $x = 8$

7. $x = -6$

8. $y = \dfrac{7}{6}$

9. a

10. $y = 0$

11. $x = \dfrac{7}{20}$

12. $y = -\dfrac{7}{2}$

13. $y = \dfrac{29}{7}$

14. b

15. $x = \dfrac{1}{8}$

16. $x = \dfrac{4}{5}$

17. $x = -2$

18. $x = 6$

19. $x = 12$

20. b

21. e

22. b

23. $y = \dfrac{e - ax - cd}{b}$

24. e

25. e

26. $l = \dfrac{P - 2w}{2}$

27. $r = \sqrt{\dfrac{A}{\pi}}$

28. $x = \sqrt{\dfrac{d + bc}{a}}$

Chapter 17
Verbal Problems

1. $x = 2$

2. $3, 9$

3. 20

4. 3000

5. $6\dfrac{1}{3}x = 114$

6. $45,000

7. $51.50

8. 63,000

9. $16,000

10. 19

11. $66,500

12. $416.67

13. 951

14. 90

15. c

16. b

17. c

18. b

19. b

Chapter 18
Inequalities

1. $x < 6$

2. $x > 7$

3. $x > 5$

4. $x > 8$
5. $x \geq -8$
6. $x < 5$
7. $x \geq -3$
8. c
9. $x \geq 5$
10. $x \leq -2$
11. $x > 12$
12. $x > -8$
13. $x \leq 3$

Chapter 19
Factoring

1. $3x(x - 3)(x + 1)$
2. $4(x + 3y)(x - 3y)$
3. $(x + 4)(x - 3)$
4. $4xy(2x^5 y^6 - xy^2 + 3)$
5. $(x - 5)(x - 2)$
6. $3x(x - 4)(x - 3)$
7. $7x^2 y(-y^2 + 2x^3)$ or $7x^2 y(y^2 - 2x^3)$
8. $(x - 2y)(x + 2y)$
9. $4(x - 4)(x + 4)$
10. $3x(x - 2)(x - 1)$
11. $(3x - 4yz)(3x + 4yz)$
12. $(x - 2)(x - 1)$
13. $(x + 9)(x - 1)$
14. $(x^2 + y^2)(x + y)(x - y)$
15. $(4A^2 + B^4)(2A + B^2)(2A - B^2)$
16. a
17. c

Chapter 20
Rational Expressions

1. $\dfrac{x - 2}{x + 4}$
2. a
3. $x - 4$
4. $\dfrac{1}{x + 5}$
5. $\dfrac{x - 4}{x + 2}$
6. $\dfrac{A + 5}{A + 2}$
7. 3
8. $\dfrac{x(2x + 1)}{4(x + 1)}$
9. $\dfrac{x}{3(x - 2)}$

10. $\dfrac{x + 3}{x - 4}$
11. 1
12. $\dfrac{x}{x + 3}$
13. $\dfrac{x + 2}{x + 3}$
14. c
15. $\dfrac{A^2 - A - 10}{2(A + 5)(A - 5)}$
16. $\dfrac{5x - 8}{(x + 4)(x - 4)(x + 4)}$
17. $\dfrac{y^2 - 8y + 18}{(y - 4)(y - 3)}$
18. $\dfrac{2x^2 - 4x}{(x + 6)(x - 6)}$
19. $\dfrac{x - 4}{2(x + 6)}$
20. $\dfrac{1}{5}$
21. $\dfrac{-x^2 + 16x - 36}{(x - 9)(x + 9)}$

Chapter 21
Quadratic Equations

1. 5, 3
2. 9, 4
3. 5, –2
4. –12, –1
5. 4, 2
6. b
7. 4
8. –7
9. 15
10. –8
11. c
12. $k = 12$

Chapter 22
Simultaneous Equations

1. e
2. (2, 2)
3. (3, 0)
4. (–1, 4)
5. (4, 0)
6. (3, 6)
7. (2, 1)
8. (3, 0)
9. $A = 0, \ B = 1$
10. (2, 2)
11. $x = 2$

Chapter 23
Graph of a Straight Line

1. $m = 1$
2. $(3, 0)$
3. $(0, -2)$
4. b
5. $m = \dfrac{3}{7}$
6. a
7. b
8. a
9. e
10. a
11. a
12. $x = -\dfrac{3}{2}$
13. a
14. d
15. $y = -3x - 18$
16. $y = \dfrac{1}{4}x + 9$
17. $y = -2x - 3$
18. $m = \dfrac{5}{3}$
19. $y = -\dfrac{3}{2}x + 3$
20. a
21. d
22. $m = -\dfrac{8}{11}$

Chapter 24
Radicals

1. 8
2. $2x^{11}\sqrt{2x}$
3. a
4. 0
5. $6x^9 y^{10}\sqrt{xy}$
6. $4x^2 y^4\sqrt{3x}$
7. a
8. b
9. e
10. b
11. c
12. a
13. $8 + 3\sqrt{7}$
14. $-9 - 4\sqrt{5}$
15. a
16. d
17. d
18. $3xy\sqrt{x} + 8xy\sqrt{y}$
19. $-9\sqrt{21}$
20. $x + 6$
21. $w = 137\dfrac{2}{7}$
22. $9x^8 y^9$
23. $40x\sqrt{3y} + 2x\sqrt{5y}$
24. $\sqrt{3}$
25. $x = 0$

Chapter 25
An Introduction to Geometry

1. $x = 10$
2. b
3. $6x^3 y^3$ square yards
4. $4x^2 y^2\sqrt{y}$ feet
5. $8x^2 y^2 + 6xy$ feet
6. $2w^2 + 6w$ square feet
7. $6x^2 y^2$ feet
8. $4l - 4$ feet
9. $32A$ yards
10. $625x^2 y^2$ square feet
11. $2\sqrt{3}$
12. 12
13. $4x$ centimeters
14. $6xy\pi$ centimeters
15. $9x^2\pi$ square centimeters
16. 12 centimeters

Algebra Assessment: Practice Placement Exams

Exam A

1. b
2. e
3. a
4. e
5. d
6. c
7. b
8. a
9. b
10. a
11. d
12. b

Exam B

1. a
2. e
3. d
4. c
5. a
6. d
7. b
8. d
9. c
10. a
11. e
12. e

Exam C

1. b
2. d
3. d
4. d
5. a
6. c
7. d
8. d
9. d
10. e
11. c

Exam D

1. a
2. d
3. d
4. a
5. e
6. d
7. d
8. e
9. c
10. c
11. a
12. b

Exam E

1. e
2. e
3. d
4. b
5. b
6. e
7. d
8. c
9. e
10. a
11. b
12. e